ISBN 978-1-332-75594-3
PIBN 10284396

FB &c Ltd, Dalton House, 60 Windsor Avenue, London, SW19 2RR.
Company number 08720141. Registered in England and Wales.

For support please visit www.forgottenbooks.com

AMERICAN FOUNDRY PRACTICE.

TREATING OF

LOAM, DRY SAND AND GREEN SAND MOULDING,

AND CONTAINING

A PRACTICAL TREATISE UPON THE MANAGEMENT OF CUPOLAS AND THE MELTING OF IRON.

BY

THOMAS D. WEST,

PRACTICAL MOULDER AND FOUNDRY MANAGER; MEMBER OF AMERICAN SOCIETY OF MECHANICAL ENGINEERS, AMERICAN FOUNDRYMEN'S ASSOCIATION, CIVIL ENGINEERS' CLUB OF CLEVELAND, AND HONORARY MEMBER OF THE FOUNDRYMEN'S ASSOCIATIONS OF PHILADELPHIA AND CHICAGO, AUTHOR OF "MOULDER'S TEXT-BOOK" AND "METALLURGY OF CAST IRON."

Fully Illustrated.

TENTH EDITION

FIRST THOUSAND.

NEW YORK:
JOHN WILEY & SONS.
LONDON: CHAPMAN & HALL, LIMITED.
1900.

PREFACE TO FIRST EDITION.

In offering this book to the public, the author would state that he has tried to select such matter as would illustrate the varied workings of difficult castings, and to offer problems for thought and study to PRACTICAL MOULDERS, in which he has endeavored to make everything PLAIN and PRACTICAL, so that the beginner or apprentice can understand it as well as the practical moulder. *The illustrations* shown are from drawings made by the author, and embrace almost all the more difficult kinds of heavy castings. They are chosen because they involve some of the highest elements of the art of iron moulding. Pattern makers and foundry managers, in considering the best method of making difficult castings, can refer to these pages, where it is hoped they will find assistance of such a nature as will often assist them in the manipulation of intricate jobs.

The class of moulds taken for illustration not only describes just how each special job can be *well moulded*, but also gives ideas which may be applied in a hundred and one other forms and to various classes of work. Also, the modes presented for doing work will be acknowledged as *mechanical* and *progressive*, and such as in many cases present original ideas of practical value for assisting the moulder to successfully produce *good*, *sound*, *clean castings*.

The melting of iron is a subject which will be found quite condensed and simple in its treatment, although it is of great importance; and from the ample illustrated workings of the foundry cupola and its management, it is believed many valuable and practical ideas will be derived.

All the matter here collected is the result of many years' experience and practice, not only as a workingman alone, but also as a manager of foundries. The author, having traveled over and been employed in different sections of this country, has had an opportunity of obtaining a varied practical knowledge of the AMERICAN FOUNDRY PRACTICE.

The ORIGINAL articles here submitted have, to some extent, appeared in the *American Machinist*; but they have been expanded, and in some cases rewritten for this book, in hopes that the minds of practical men may give thought to the subject, and that other tradesmen will learn that the moulder's trade is one that requires *something higher* than the mere muscular force necessary to pound sand.

The field for *thought* and *study* in foundry practice is very large; and if the author, in presenting these pages before the PRACTICAL MACHINERY MOULDERS OF AMERICA, has benefited a class in whom he takes pride as a member and co-worker, he will feel amply repaid for his labors.

THOS. D. WEST.

CLEVELAND, *September*, 1882.

PREFACE TO SIXTH EDITION.

It will be seen by comparing this edition with previous ones, that a thorough revision of the whole book has been made, so as to not only keep it up with whatever progress there has been wrought in the branches of moulding of which it treats, but also to have it as a handbook to the author's second work, "MOULDERS' TEXT-BOOK," contents of which are given at end of this book.

The sales of both works have exceeded the author's most sanguine expectations; and the commendations they receive from their readers make the author feel recompensed for the arduous study and labor spent in compiling the works.

Thanking the many that are so kindly advocating their value and introduction to the trade,

Yours truly,

THOS. D. WEST.

CLEVELAND, O., *May*, 1887.

PREFACE TO TENTH EDITION.

ALTHOUGH it is seventeen years since the first edition of this work appeared, the sales show as great a demand to-day as during the first few years of its issue. This is no more than the author anticipated, for the subjects treated are such as set forth the principles which underlie the art of making castings, so little known seventeen years ago. And they have been presented in a manner which it is hoped all may understand, and further may apply them successfully. Since the first edition, every week has brought letters to the author sustaining the belief that his methods have been appreciated, and this to a large degree has encouraged him to continue his researches in problems of practical founding.

The author's interest in the advancement of his trade has caused him to bring out two other works, both of which have had a good sale, which is the best recommendation of their utility. While these later works necessarily treat many subjects in a manner similar to this book, they give more knowledge of an advanced character, due to the author's extended experience and researches. Hence in order that all the subject-matter of this book shall be as modern as that of the author's two later works, it has been found necessary to revise this volume and to add new matter to some few chapters. It will therefore be seen that a thorough revision of the whole work has been made, and for this the author is much indebted to the good will and kindness of his publishers; and he trusts that the sales in years to come will, as in the past, remunerate them for their outlay and for the faithful manner in which they have handled his works.

THOS. D. WEST.

SHARPSVILLE, PA., *June*, 1900.

CONTENTS.

THE MOULDER AND THE FOUNDRY.

GREEN SAND MOULDING.

LOAM AND DRY SAND MOULDING.

MANIPULATION OF IRON CASTINGS.

NOTES AND RECEIPTS.

INTRODUCTION.

THE MOULDER AND THE FOUNDRY.

THE MOULDER AND HIS TRADE.

ASK any mechanic what trade he thinks requires the greatest amount of mechanical ability and he will say his is the one, and perhaps go on to state some of the fine points connected with it. If the moulder should be asked this question, he would probably get excited over it, on account of the low estimation in which his trade is held.*

The moulder's trade may not be the most mechanical of all trades, but it is decidedly entitled to more respect and consideration than is usually given it. Other tradesmen must remember that to be a good moulder requires more than the muscular force necessary for ramming sand—an idea that has been expressed time and again. The machinist with a clean Monday suit on and a pair of calipers in his hand; the pattern-maker with his plug hat looking over his drawings; the blacksmith making the sparks fly; all have a dignified appearance. The position of a moulder lying on his back under a cope, or on his belly ramming under some pattern, is not suggestive of dignity. The general impression is that the nicer the clothes, the more mechanical is the trade.

The moulder with his black face and clothes, and surrounded by the usual appliances of a foundry, such as

* This was in 1882, but the distribution of practical literature on the science of founding has since greatly elevated it in the estimation of all artisans, engineers, and scientists.

bricks, loam, mud, ashes, straw, horse manure, blacking, sand and clay wash, might have a romantic, but is far from having a dignified appearance. Should he attempt to put on dignity, when he is prostrated or laid out for repairs on some sand heap, caused by carrying hot iron or doing a heavy feeding job, it would be all knocked out of him. Like a man picking up hot iron, he would be forced to lay it down again.

It is this want of dignity about a foundry that lowers the trade in the estimation of many, and the moulder will have to look for other things than dignity to establish respect for his trade. In many ways we moulders are as a class to blame for the disrespect generally shown to our trade. In the first place, too many have no respect for themselves; and, in the second place, too few study to *understand the principles involved in the moulders' art*, or do anything to compel their recognition by the mechanical world.

Our trade is one in which there is certainly display for mechanical ability. In fact, the author knows there is very great mechanical ability required to make a man a *reliable, good moulder*. To prove this to the world, and thereby command respect for our trade, should be the aim of every moulder.

How sublime and grand is the structure of the steam engine—the mighty power of machinery! The useful and ornamental castings used in all shapes and forms are made from pig iron and old scrap iron, and formed in sand. For all this, all thought is of the work of the pattern-maker or the machinist. It is not till some scabs or sand-holes in the casting are noticed that the moulder is mentioned. Then what abuse the poor moulder does get!

All moulders are not thorough-paced; if they were, there would not be half the trouble there is in getting good castings. There is no trade that requires more long-headed, cautious and mechanical operations than that of the moulder.

Why is it that all the castings made in a foundry cannot be good and perfect like the day's work of a machine, blacksmith, or boiler shop? Is it because the men and boys that learn moulding are such as are rejected or not allowed to learn other trades on account of being blockheads? If it were possible that such was the case, it would then be reasonable to say that should any other set of tradesmen have learned moulding, bad castings would never be seen. The moulder's trade is learned by boys and men, the same as any other trade, and foundry bosses are as good judges of character as any other class of foremen. Few foundries would hire a boy that wore kid gloves and a collar that holds his head up. The boys are generally selected from sound and staunch material. Taking it for granted that as good and as smart boys learn moulding as learn other trades, is it to be taken for granted that they fail as a class when they become men? The more we think of the matter, the more it looks as if it was the want of a knowledge of moulding more than the lack of mechanical ability that causes all this trouble of bad casting. This article is not written to hide the moulder's failings, but to get at the truth, no matter where or whom it strikes. The moulder should admit that, when he loses a casting which he has *had full control of, and proper tools and materials to work with*, it is no more nor less than his ignorance or carelessness that caused the loss. The proof is that when he makes it the second time he gets a good one. The loss of a casting does not imply that the moulder is ignorant or is not a mechanic, since castings are often lost from some little, insignificant cause. There are a thousand-and-one ways of losing castings; and the moulder, when making the second casting, is nearly as liable to have that bad, from some other cause; and the moulder does not live who never lost a casting.

Sometimes the excuses for bad castings are laughable.

The story is told of a moulder who made four pieces—every one bad—and, when the foreman asked him what was the matter, he said that one dropped, one flopped, one run out, and that one was a "waster." The boss told him to make one more, as he would like to know what would be the matter with the fifth one. Ask any moulder if the bad casting which he has made cannot be made good, and why it is bad, and he will answer the first question in the affirmative, and have some excuse, instead of an answer, to the last one.

When a moulder loses a casting, it worries him. There is no trade in which a man's peace of mind is kept so unsettled as in the moulder's. He is always in a state of expectancy. Look at a moulder when he is taking his casting out of the bricks or sand, and with a hammer in his hand, he will look for something that he does not want to find. Should anything be seen that would make the casting bad, how soon the honest man's look of fear changes to despondency, or he shows his character by throwing the hammer down and stalking around the shop with a look of indifference, as much as to say that he was not responsible ; or he will seek consolation by laying the blame on some poor helper core-maker, or on some moulder that worked with him. It takes a moulder that is a sweet talker to get out of the blame for a bad casting, when he knows there was no one to blame but himself. Losing castings with one moulder is a frequent occurrence, while another will be noted for success. This success may continue a long time, on the strength of which he will get careless, and some day, to his sorrow, he has a bad casting. He makes it over again, guarding with the greatest of care the conditions that caused the first one to be bad, and his mind being riveted to this point, he neglects others, and the second casting goes the way of the first. If now he has not a well-balanced mind he may lose almost every casting he tries to make, and it is not till he makes an effort to overcome his nervousness and lack of confidence, that he will be able to make a reliable mould.

A casting made by a half-drunken moulder would be more likely to be good than one made by a nervous moulder.

Any moulder when starting on a large responsible mould should have a clear head, so as to master the job with his brains before he puts his muscular forces to work. This will give him confidence, which along with a good mechanical judgment is a very essential feature in making good castings. About the best proof that moulding is a trade that requires the best of physical and mental power, is to notice the moulder when his castings come out all right, and likewise when they are bad. In looking at a bad casting, the question is always asked, What made it bad? Such a question implies that the cause is not apparent, but that it needs investigation.

The blacksmith when forging his iron into any shape with his hammer, can, the same as the machinist or pattern-maker, see the effect of every movement he makes as being a move towards the end. Should any part not be done right, it will be visible to the eye, and the little mistakes can be remedied without waiting till the whole job is completed.

Moulding is like to a man fishing, he cannot see what he will get until it is out of the water; and he may spend all day working hard to catch something, which when brought to light will be a worthless minnow.

The ramming of sand is what any one having the necessary strength can do; but the light or heavy ramming required on the different sections of a mould, demands something more than strength and stupidness. The *motion* of the rammer is visible, the *result* of the ramming is invisible.

A moulder may work from one day up to one or two months on a job, and every night when he goes home he feels anxious to know the result of his day's work. There are often times when a moulder would forfeit his day's wages if he could only see or know the result.

Often things happen to castings that will puzzle the best of moulders to fathom, and which, when found out, involve some chemical or scientific principle that learned men are very proud in talking about.

LEARNING THE MOULDER'S TRADE.

When a young man starts out to learn the moulder's trade, about the first thing he does is to get a trowel, stick it in his pocket, and call himself a moulder. He comes to his work finely dressed, with a cigar in his mouth, and his talk is about anything rather than what he is doing. This is not the case with all beginners, but it is true of the majority of them. Once in a while a young man who has more *sense* and less *conceit,* instead of calling himself a moulder takes every opportunity to make himself one. He is careful not to give offense by speaking slightingly of work he knows nothing of, and at once makes friends of those whose skill he may profit by. He may wear good clothes, and perhaps smoke a cigar, but there are different ways of doing such things. Instead of spending his evenings around saloons, he may be found a member of some society, studying to improve his store of knowledge, or he is home making a drawing representing the way some moulder is making a difficult casting, which drawing he will preserve for future reference. When a moulder loses a casting he will note the cause and profit thereby, and when he loses a casting himself, he will welcome and profit by any advice or assistance in order that the next one may be good. At his work he will be diligent and careful, and always ready to give a "lift" or help any one that is in trouble. He is not afraid of asking questions, and always aims to make the second casting better than the first one. He will be patient, and not be looking for the foreman to give him work that he is not capable of executing. When he borrows a tool it is sure to be returned, and his own tools he willingly lends and keeps them clean and in place. He never has much to say, and attends strictly to his own business.

These qualifications in a young man will make friends, without which his progress will be slow. The greater part of our knowledge is obtained from others. We are indebted to thousands of people for what we know of the moulder's trade.

Some beginners will say that they do not get any show; that the boss is giving it all to others. It is hard for an outsider to pass an opinion on this. Very often apprentices overrate their ability, thinking they are capable of taking work that they would only lose if given them to do. It is a great failing of young men, and, in fact, of the human race generally, to think they can do things they see others doing. If we could only "see ourselves as others see us," we would, in many cases, be more contented in our situations.

It is often the case that a worthy apprentice is not advanced as he should be, on account of some prejudice, or, sometimes, an established principle of keeping the boy down. A young man should find out the character of the establishment before he makes an agreement to learn the trade there. A shop that makes a specialty of two or three different kinds of casting, is no place to learn the moulder's trade. Try and get a start in a good jobbing or steam engine foundry; such shops as these generally have all the science of the art of moulding practiced in them, and in such shops, should the foreman feel inclined to keep a beginner on one job all the time, he could very seldom do so. Of course he can keep changing you from one inferior job to another, and should you see all the apprentices treated in a similar manner, it is not of much use to ask to be advanced. But, should you see others going ahead, (if you are sure the fault is not yours) in a polite manner ask to be tried on better work. Whether the answer is favorable or not, continue on with your work, doing the best you can, for, if the foreman will not reward any merit that you may possess, you will be noticed by some one, sooner or later, who will recommend and advance you.

Under any circumstances, faithfully serve your time, then leave and go to some other shop, wherever you can get a job.

You may find quite different plans for making work practiced there ; but, with a good mechanical judgment, you will get along all right. The first day in a new shop is always the worst. I have seen men, who have worked the most of their lives in the shop where they served their time, and in which they had the leading work, and were reckoned good mechanics, start in a strange shop and be so nervous and simple in their actions that the old hands would question their being moulders.

As a general thing, the class of men who laugh at a moulder in a strange shop are the narrow-minded ones who, having had experience several times over with every piece made, have forgotten their own failures in working up to their present knowledge. A man of good sense, and who is a *thorough mechanic,* will not be guilty of such actions. On the contrary he will show the stranger where he will find the flask needed, and will tell him if there has been any trouble in the previous moulding of the job; he will show him where there are gaggers hidden, and, in fact, do everything that he can to assist him. Should he have any idea of opposition, he will wait until the stranger has got a fair run of the shop's tools and ways, when it would be a more manly and even race to see who is the best mechanic.

A thorough knowledge of the moulder's trade cannot be learned in any one shop, nor is it a sign that a moulder thoroughly understands his trade because he has worked in a great many foundries. He can see how things are done by traveling, but the class of work that would advance and instruct him is hardly ever given him to make. Go into any foundry and ask how long the men that are working on good jobs have been there, and the answer will generally be, from six months up to a lifetime. A stranger must stay long enough in a shop to show some merit before a practical foreman will trust him with responsible work.

A young man traveling to advance himself should, when possible, engage only in the best shops to be found, and there he should stay at least for one year. After thus working for ten years in as many different shops, he can blame no one but himself if he is not a good, practical moulder, able to make almost anything in the branches that he has practiced.

Thorough, first-class moulders are very scarce, as such men must be capable of melting their own iron, and making any castings that come along, in loam, dry sand, or green sand.

It is very seldom that the three branches are learned, or practiced, by one man, one reason being that most large shops generally have work enough to keep a constant number of men working steadily in each of the three branches. Another reason is that it is a little too much for most men to practically master.

A man may be good on green sand, and perhaps fair on dry sand and loam, or all right on loam and dry sand, yet in green sand not amount to much.

There are two ways of learning the moulder's trade; one is *do as you see others do,* and the other is to know the reason, *why you do so.* Moulders very seldom ask themselves: Why is such a thing done in order to have the casting a good one? They are told it is to be done; they do it and there let the matter end.

There is a principle and a cause involved in almost everything that is done to make castings successfully, and he is the farthest advanced in the art of moulding who has made them the study, so as to thoroughly understand the cause and effect of what he does.

It has been suggested to me that I should write a few articles for apprentices. Webster says that an apprentice is one that is bound to another to learn a trade. Some trades may be learned during the allotted time of three or four years, but for a young man to think that when his apprenticeship is served he has learned the moulder's trade is assuming too much. A moulder is an apprentice as long as he lives, as there is not a day that passes that something cannot be learned. Whenever any man gets to thinking that he knows it all, or that he cannot learn any more, he should stop working. He will never be a success. Writing to give information to a beginner sometimes works very well, but the beginner must first see some bad results of something that he has done, in order to fully understand the importance of the instruction previously imparted to him.

The first year of a beginner's time is always more or less of a loss to his employers. You may tell him what to do and how to do it, but he must have practice before, as a general thing, any information that may be given to him is fully understood, or its value comprehended. Articles are often written for apprentices, when the author ought really to admit that he intended them more for what might be called practical men, and that he assumed the simpler title to cut off censure and criticism. Such authors should let their writings be for the old as well as the young, for there are none of us so old that we cannot learn.

A beginner in one shop could very often give some valuable information to an old experienced hand in another shop, and as for a knowledge of the principles or manipulations of the moulder's trade, there are as many old hands as new ones that require to understand them better.

It would seem to many that with all the advance made up to the time of this revision (1899) in reducing the cost of castings and obtaining more perfect work, the skill of the moulder should have advanced accordingly. The author regrets to say that such is not the case. We have far fewer thorough moulders to-day, in proportion to the number engaged in the trade, than existed fifteen to thirty years ago. This is due to the fact that, in years gone by, the moulder, in order to find and hold a situation, had to be able to make almost any kind of a casting in either loam, dry or green sand work, according as he was experienced in either or all of the three branches. Employers and journeymen, realizing this condition, influenced the beginner to serve long terms of apprenticeship, which as a rule gave him a constantly changing practice. There are to-day but few shops, comparatively speaking, that can give an apprentice anything like a wide range of work in either loam, dry or green sand moulding from which to gain experience. The majority of founders to-day, being manufacturers of certain specialties, can, at the best, but let a boy pass through the limited range of work in their own shops, which is almost as nothing in many cases, compared with the experience a boy could gain in many shops years ago. It is true we still have some good shops in which the boy can obtain the foundation for making a thorough moulder, but as a rule the boy that does get this golden opportunity now rarely appreciates it, and his buoyant Americanism is apt to cause him to disrespect the restraint of apprenticeship service, and often impels him to kick over the traces and leave a shop before his intended apprenticeship is half served, and not until years after does he discover the mistake he has made, when he is finally ready to make every effort to perfect himself in a knowledge and experience of the trade. It is the above condition of affairs that is largely responsible for many men's interest in practical literature at the present day. The author is in a position to know whereof he speaks in this matter, and from his experience in being in touch with readers of books and writings on founding, he can say that to-day's practical literature is in many cases perfecting the moulder or founder more than shop experiences. Take away all literature that teaches cause and effect and the principles underlying the art of founding, as taught by the author's works and writings of similar character, leaving nothing but the specialty shop and curtailed apprenticeships that often give experiences little other than to pound sand, and we would find our trade in a sorry plight, without any men worthy the name of moulders. There are many shops at the present day (1899) suffering keenly the need of greater skill in the moulder, and the wise beginner or aspirant to promotion of skill in the trade will not leave a stone unturned in laboring to perfect himself as far as he can. There is plenty of room at the top for the molder to command for himself wealth and positions of trust sufficient to satisfy the ambition of almost any true mechanic, and making a study of the principles underlying cause and effect in founding, such as the author's works define, will greatly assist any attaining such heights.

BUILDING A FOUNDRY.

WHEN a man is about to construct a foundry, he cannot give the matter too close attention. Let him make lines and rub them out again until he gets something that fills his ideas; then make three or four tracings, and submit them to as many different practical foundrymen, with the request that they find all the fault with them they possibly can. Then, with a mind unprejudiced, let him consider their opinions, and adopt whatever is good. This plan, if adopted by men not experienced in foundry practice, should result in giving them a well-arranged shop.

The idea that should be prominent is that the plan of a foundry should be decided upon from a consideration of the particular class of work for which it is to be used, and other controlling circumstances, such as the general character of the land, the position of a railroad, river, lakes, streets, etc., etc.

To attempt to show a plan for the construction of a foundry that should be of anything like general use in building would be foolish, since scarcely any two foundries ought to be built alike. The fact that there are so many unhandy foundries is not always evidence that the designer was in fault; since, considering the location of other buildings, and circumstances over which he had no control, he may have done the best that could have been done.

There is one thing, however, builders or designers are greatly to be blamed for, and that is for not providing for enlargement.

If there is not much business or capital to start with, the small shop, having only one cupola, one crane, one pit, and one oven, is not the building that should be made on paper. The proprietor or designer should take into consideration his available ground-room, and then make a drawing or plan as if he were going to build the largest shop that could possibly be constructed on the grounds. If there is room for three or four cupolas, a couple of air-furnaces, five or six cranes, a number of different-sized pits, and several good ovens, let them all be carefully located on the large plan or drawing. When the drawing is completed, let him consider what portion of his large shop would be the best and cheapest for him to construct, with the capital he can afford to invest in his enterprise to start with. Then, when his business increases, and he wants another crane, cupola, pit, or oven, he will only have to look at his original drawings, and there are places for them. When he builds his shop larger, the builder can find studdings, bolts, or broken brick-work to securely fasten the extension to.

It is not intended that the reader shall take the word "extension" to mean the usual kind of extensions that are added to foundries, such as "dog houses," "pigeon holes," etc., and which, wherever seen attached to the main shop, are sure signs that extension was never thought of, or provided for, when the main building was first planned.

There is no intention in this to show how to build fine, large shops, but rather to show to the man of small capital that before he starts to lay out his money he may, to a great extent, by careful study and management, make his little enterprise a running success. Many a man has failed for want of judgment in the beginning.

There are two things that are connected with every enterprise. One is the advantage and the other the disadvantage. When a man does not see *both*, it is evidence that he has not deeply investigated the subject. A man who takes every element and business point separately, and thoroughly dissects them, not only can know what is best for him to do, but will be inspired with such confidence and energy that the word "failure" would have to be printed in larger type than is yet used for him to see it.

In building a foundry, the shop should be built high. The medium height for shops that do crane-work is about twenty feet. This measurement is from the floor to the large girders, or beams, that the top of the crane is held by. In fact, any foundry should be built high, so as to give plenty of space for the gas, smoke, and steam (which is always generated at casting time) to rise up over the men's heads. To carry hot iron through a dense fog of gas, smoke, and steam is a duty that is not only unpleasant, but has been the means of many workmen getting badly burned.

The next point, and one of great importance, is to have the shop constructed so that plenty of light will be admitted from the roof, as well as from the sides. A dark foundry is not only disagreeable to work in, but is the cause of many rough and poor castings. It is also a great drawback in getting out work fast.*

A foundry that is built for large, heavy work, cannot be too strong. The doors, or openings, through which the large castings are delivered, should not be less than fourteen feet wide and ten feet high. It is best to have the doors hung by weights, so they will slide up and down. Doors that open out or in, or that run backward and forward on sheaves, are always more or less in the way. Doors, when it is possible, should be placed in a part of the shop so that when opened the dust cannot be blown on the moulder

* While it is necessary to have good light, too many windows will cause the sun's rays to be injurious by drying out sand-heaps and moulds, which makes trouble in finishing and may cause bad work.

or his mould, as it is not only disagreeable but it hinders him from doing his work.

Cupolas should be built in that part of the shop in which the large doors are situated, as generally this part of a shop is only used for a road or gangway. Placing doors and cupolas near together utilizes room, as the room for several feet around a cupola is not used for moulding. This plan also keeps the dirt and dust from the doors and cupola together, and the moulding room, destroyed for one, will answer the purpose of the other.

Loam and dry sand moulding should be kept in a part of the shop distinct and away from the green sand floors or moulding room, as the dirt and mess that pieces of brick, mud, straw, cinders, etc., make are very disagreeable, and a hindrance to the green sand moulder. The best part of the shop for loam and dry sand work is at one end, and near the ovens.

The ovens should be located in the part of the shop where there is not much traveling done, either by cranes and cartage, or foot travel; also where the railway tracks that the oven carriages run in and out on, will take up the least valuable room.

Large and small pits for casting, or ramming up moulds in, should be as handy and as near as possible to the loam-work.

When air furnaces are required, they should be located as near as they can be to the loam-work, and where there will be nothing in the way of delivering heavy or large scrap iron to charge them up with. They should be built up enough above the level of the foundry floor so that the tapping hole will be from three to four feet above the floor, in order to admit the pouring of moulds direct from the furnace, or to have the liquid iron first run into a large basin or ladle, from which it is admitted into the mould.

Every well-regulated foundry should have good facilities for cleaning castings, which, when possible, should be cleaned in an adjoining room, so that the moulders will not be hindered from their work by waiting for a crane, looking out for flying iron chippings, and giving orders for hoisting and lowering the crane, that cannot be heard on account of the noise.

A narrow or wide track can be laid between the casting and cleaning shops, and as soon as the heavy castings are hoisted out of their moulds, they can be loaded on a car and run into the cleaning department, in which there should be a crane for handling them.* In the cleaning room there can be tumbling-barrels and vitriol-tubs for the cleaning of small castings.

Shops that do heavy and light work should have the light work done in parts of the shop entirely separated from the heavy floors, for the reason that grades of sand better adapted for each class of work can then be used, and the work done to pay better. The portion of the building to be used for the moulding of heavy castings should be constructed with a view to strength, while the portion for the light castings can be constructed more cheaply.

In selecting ground to build on, there should be three or four wells or holes dug to see if it is subject to dampness or water. Should there be water found at the depth of six to eight feet, the position should be rejected—that is, if the foundry is to be constructed for a heavy line of casting that will require bedding in the floor, or should pits be required.

When planning a shop, there should be plenty of time taken before it is let pass into the builder's hand for construction. Hasty planning is likely to be sooner or later regretted.

---

* A plan of a very useful car is shown on p. 231, Vol. II.

The old plan of building foundries was to construct them of frame or brick. It is a rare thing at this time to see a wooden shop built for this purpose; and, as a rule, only proprietors having small means, build wooden structures. The modern plan is to build either of iron or of brick and iron. The iron shop has not proved satisfactory, as the sheeting rusts out in a few years, and causes the roof to leak; it also allows the chilly blasts of winter to enter the shop through its sides. These, of course, can be repaired, but, as a rule, the result is annoyance and often bad work before this can be done. Then, again, frequent repairs become very expensive, for, as a general thing, the sheeting must be painted inside and out every two years. It is far better to make the sides of brick and the roofing of iron, so that it can be covered with slate instead of iron or steel sheeting. A shop built on this plan may last for fifty years without giving much trouble from leakage due to bad weather, and in the mean time costs but little for repairs, compared to an iron structure. It is far better at the start to build the best, and it will pay well in the end. In 1882, when this book was written, we had, as a rule, foundries poorly built and badly equipped, but 1899 finds us with good shops which are not only economical but convenient.

Green Sand Moulding

MOULDING AND CASTING FLY-WHEELS.

The engravings herewith shown represent different plans of sweeping and moulding fly-wheels. The lower cut is a wheel that had a full pattern for the arms and hub, the rim being swept out. For forming the cope part of the rim there were wooden segments, *R*, used. A straight sweep which formed the joint, also struck or marked on the joint a true circle to set the segments by. The reasons for using the wooden segments was that the wheel was quite a heavy one, and if the moulder did not gagger it well the cope would be likely to draw down.

When gaggers are set on the sand-bed or mound, they will generally show their prints, so as to require knocking back, which, when there is a large cope surface, or a large numbers of gaggers, requires considerable time and labor. Besides this, it is not always a safe plan to knock back gaggers, as it will generally loosen them, thereby causing trouble.

Some moulders ram up large copes without marking the mould with them, while others make their mould look as if they intended to use the hills and hollows for a guide to close the cope on by, instead of stakes or pins.

In starting to mould this wheel, a good coke or cinder bed is made, the spindle seat set, and sweep, *B*, attached. As this was a half-wheel, the half-hub and three-arm pattern was bedded in true and level, using for a guide the faces of the sweep. After the arms and half-hub were bedded in, the outside or rim was rammed up, after what some might think an odd plan. Instead of ramming the rim, or outside, up, for the purpose of forming a surface to ram up the cope on of all common sand, which has to be shoveled out when the cope is lifted off, in order to ram it up again to form the sides and bottom of the rim, the following plan was adopted. The adoption of this plan not only saved the extra work of ramming up a large hole twice, but also gave a more solid mould than could be formed where there is only the sweep to work with.

In first starting to ram up this rim surface, the sweep is set to the right position, and common sand is rammed up solid within about one inch of the bottom face of the sweep. This bed is then well vented with a large wire, after which the vent-holes are stopped up on the surface, to prevent loose sand from filling them up. Facing sand is now shoveled on up to the height of about 4″, projecting out 2″ on each side over the width of the rim, on the outside of which common sand is shoveled. The sweep having been raised up, this sand is made level all around, after which this course is rammed the same as a moulder would ram any course of sand for a heavy casting. The sweep is again raised up, and another course of facing and common sand shoveled around it, the rim being all facing sand, with the common sand outside for a backing. This is repeated until the hole is rammed up high enough for a straight sweep, which is now screwed to the piece of boiler plate *P*, for sweeping off the joints.

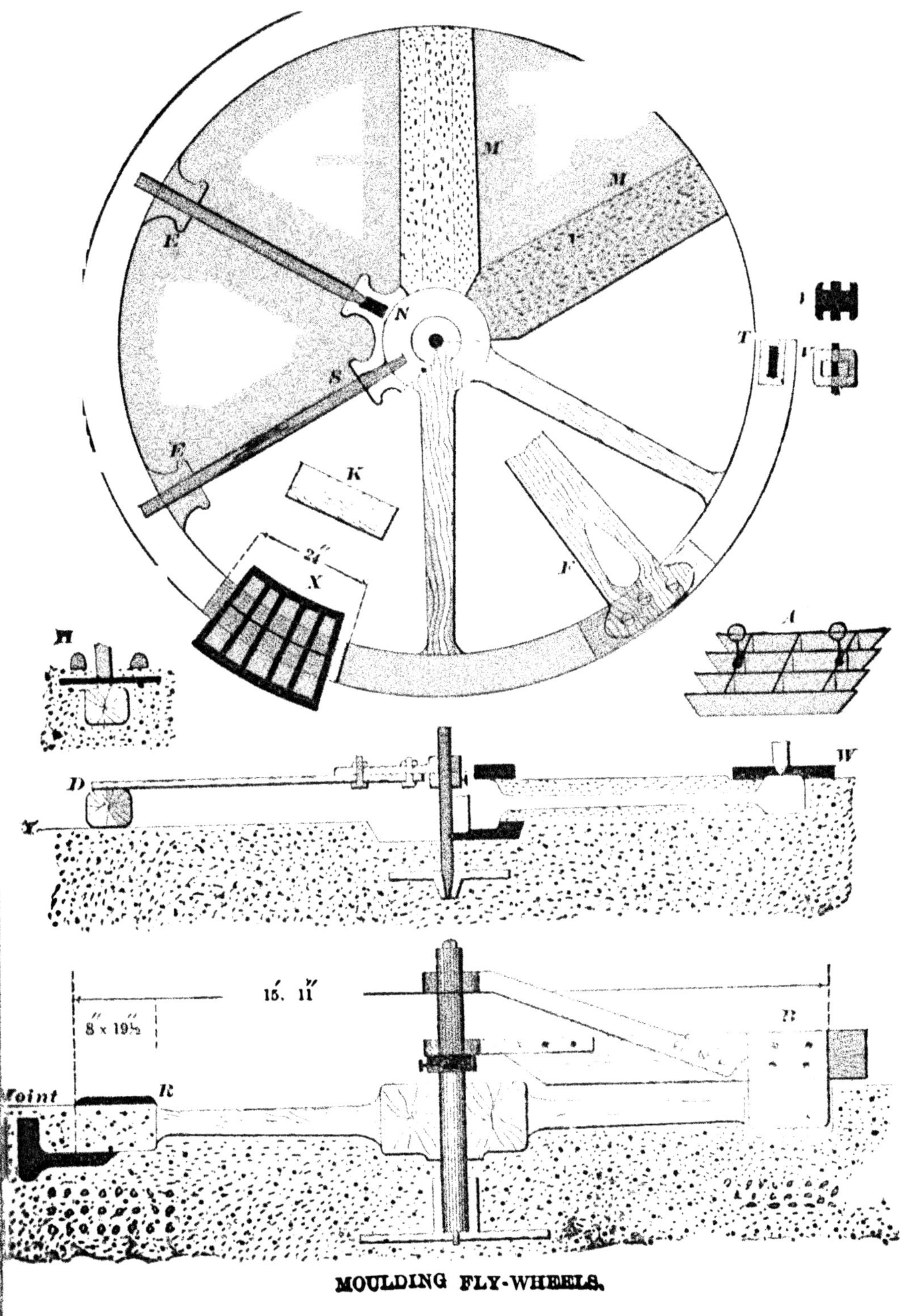

MOULDING FLY-WHEELS.

After the joints are made and the cope rammed up and lifted off, the rim sweep is set back again and the rim of the wheel is then swept as follows :

The sand is first shoveled out within one inch of each side of the rim and within about two inches of the bottom. (This sand being all facing is wheeled to the facing boxes and used again for another wheel, or for any other heavy class of work.) The sweep is now lowered inch by inch until the rim is swept out as wanted. In order to properly divide the wheel for a half one, a long straight edge is used, having a half circle the size of the spindle cut into it, which being placed against the spindle, a half-wheel is then marked off, and wooden blocks the shape of the rim, having prints for shaped cores like *V*, *V*, are then placed on the bed and the ends of the rim rammed up. The wooden arms, end-blocks, and half-hub pattern are now drawn, and the whole mould is sleeked and finished up.

The castings are poured by gates underneath the rim, as shown.*

The second, or middle cut shows another plan of sweeping up fly-wheels. The common plan of covering over large fly-wheel rims is by using segments of cores, as shown at *W;* or sometimes loam rings or plates cast with prickers on them, and then filled and swept off with loam, are used. When the oven is too narrow to dry these covering rings or plates, they are cast in two half-circles, so as to be admitted into the oven. The plan here shown is to cover the rim with green sand as it is being rammed up, which is done as follows

The level bed, *Y*, is first made and then the segment ring rim pattern, *D*, which is attached to an arm as shown, is set on the bed. After the cores for forming the arms and hub (as shown on the opposite side) are set; the rest of the inside of the wheel is swept up. The

* The utility of underneath pouring gates is described on p. 117, Vol. II.

arm that is attached to the segment rim, shown at *F* on the plan, is then taken off, the segment rim is set back, and the outside and top of the rim is rammed up, and the segment pattern having a slight draft to it, is, when ready, drawn out around endways. The moulder can now look into the rim mould, to see that everything is all right. After this replace the pattern or segment, inserting one end into the end of the mould 2″ or 3″, and ram up and finish another segment. Repeat this operation until the whole rim is rammed up.

In ramming up the last segment, the moulder may ask how the pattern is got out. This is done by having a covering core to close up the last segment, or by having a cast-iron flask as shown at *A*, one end of which is rammed up, and then the pattern is moved back after lifting off the flask. When replaced the other end is rammed. The whole flask could be rammed up at once if there was another piece of segment pattern used. For rodding over the top of the rim, which must be done in a reliable manner, there are cast-iron frames used, as shown at *X*. After these frames are well bedded on, and about 4″ or 5″ of sand rammed over them, there are some pigs bedded on to hold it down, as shown at *H*, the pigs being put on before the pattern is drawn.

This segment pattern could be improved and made to draw out endways easier by having it cut like two wedges, as shown at *K*.

For the benefit of moulders that may be afraid to try the plan, I would say that I am acquainted with one foundry where nearly all wheels are made in this way, with good success.

The upper cut shows the process of moulding fly-wheels having wrought iron arms; also for forming cast-iron arms with cores. Having a full pattern for the arms and hub; and fiy-wheels cast one-half at a time. In moulding wheels having wrought iron arms, it is generally known that the rim is east separately and first, and that the hub is cast after the rim is cold, or has shrunk about all it will shrink. Should a wheel having wrought iron arms have the hub and rim cast at the same time, there would not only be the difference of contraction of the rim and hub to contend with, but also the expansion of the wrought-iron arms, for as soon as the mould is poured, the arms will commence to expand. In a short time the rim and hub will commence to contract. The rim being much larger than the hub, its contraction will be several times greater, and with the expansion, or non-contraction of the arms, the inexperienced moulder will be able to see what the result would be. If he does not think of it at the time he will find the next morning the rim cracked or broken.

For wheels with hubs over 12″ diameter, the arms should be made with a taper on the ends that are cast into the hub, as shown at *S*. This taper can be swaged on by the blacksmith, or turned on by the machinist. For large wheels having a hub over 24″ diameter, it is better to have the taper turned, as they will then be sure to have a true smooth taper, which the hub when it contracts will have a better chance to pull in or away from.

For very large hubs it is best to have key cores set in at the end of the arms, as shown at *N*, as they can be made to answer two purposes; the first being, that should the moulder not feel safe as regards the contraction of the hub freeing itself from the arms, the key cores can be dug out, and iron keys driven in to force the arms outward; this being done while the casting is yet red hot. When the casting is cold, should the machinist find that the arms are loose, he can drive in permanent keys to hold the hub firm and stiff. This class of hubs should always be poured with iron not too hot, and the moulding of them should be done with the greatest of caution.

Before the hubs are cast, or the center core set, the rim should be tested with a pair of trammels, or with a sweep attached to the spindle, to see if the wheel has contracted evenly all around. I have seen wheels drawn out of true so much, that in setting the center core in the hub, it had to be set over ½" out of the center, in order that the hole could be bored true with the rim.

When this class of wheels are moulded without having a full pattern to work with, the rim can be swept out, or formed with a segment, and the projections *E E* can be formed in cores; but for moulding the hub it is better to have a full pattern to work with.

At *M M*, is shown a plan for making cores to form cast-iron arms. The wooden arms shown are for the purpose of making or moulding wheels that are covered entirely with a cope, having the rim swept out as described and illustrated by the lower cut.

Very often large wheels are cast in halves, having chipping or planing pieces cast on, so as to allow them to be fitted together. Without experience in making such wheels, it is often found that when it comes to putting them together they will meet at the hub leaving the rim open as shown at *T*. Chipping off the hub to bring the rim together will often make the wheel out of round, to avoid which the rim should be moulded so as to project from ½" to ¾" beyond the face of the hub.

V V shows the general plan for coring and casting half fly-wheels. Sometimes for heavy wheels there are lugs cast on the inside of the rim, so that bolts can be made in connection with keyed irons.

In moulding fly-wheels there is not always the attention given that should be in regard to having the faces true and the rim an exact circle. A good moulder will take as much pride in trying to make a wheel that will run true and even, as he will to have a smooth solid casting.

SURFACE AND BOTTOM OF GREEN SAND MOULDS.

THERE is no part of a mould that requires more precaution and judgment, coupled with a knowledge of thorough practice, on the part of a moulder, to insure a first-class casting, then ramming the surface and bottom of a mould, and there is no other part of a mould that moulders have so many different ways of handling. Take, for example, almost any pattern, and give it to a moulder to bed it in the sand; after which take it to another shop, and try a second moulder (being sure that he did not know how the first man handled it in moulding), and so on until six or seven shops have been visited, and you need not be surprised that each moulder, who considers himself a good workman, has a different way of performing his work. A few will handle the job understanding why they do certain things, while the rest will follow a series of details which has been simply taught them. The latter is a class which will have many supernatural, profound, and flimsy excuses for bad work. Moulders frequently entertain the idea that the heavier the casting, the harder should be the surface of the mould, but in my practice this has proven erroneous. There are light large castings made which require the surface and bottom of the mould to be very hard, so as to resist destruction threatened by the sudden head or pressure caused by fast pouring of the molten iron. If the beds of some solid heavy castings were made so hard, the moulds would be liable to be blown all to pieces, so that instead of the solidity, or weight of a casting being a rule for the hardness of the mould, it is better to consider the time it takes to have a head or pressure on the surface of the mould during the process of pouring. It is very easy to control the hardness of the bed of a mould that can be formed with straight edges, but the process of bedding a pattern in the sand by ramming the sand under it is more difficult, and requires more time. Some moulders will take almost any pattern, and bed it in the sand by digging out a hole and shoveling in from one to two feet of loose sand. They then take the pattern and pound it down into the soft sand, until they think it solid enough. This way of bedding a pattern is a quick but very poor one, and should be forbidden, as it is in some shops that desire to insure good work. This way of bedding a pattern also causes the bottom and surface of a mould to be exactly the reverse of what they should be, for the reason that rapping down the pattern makes the surface of the mould hard, leaving the sand soft under it, so that when the iron first enters the mould it bubbles and scabs. When a heavier pressure of molten iron comes upon the mould, it will cause the soft sand below to give way more in the middle than at the outside edges, so that when the casting is taken from the sand it is apt to be both swollen and scabbed. In moulding, the under portion of a bed requires to be rammed good and solid ; the more strain to be resisted, or the heavier the casting, the more solid should this portion be rammed. If the bed is formed with straight edges, it can be rammed solid up to within three-quarters of an inch of the top, then well vented. After this the surfacing sand should be put on, and finished by rapping it down with a straight edge, or going over it lightly and evenly with a butt rammer. This surface sand should be soft, so that when the iron enters the mould it will remain still, and not bubble or boil. For coped moulds that have a large surface at the bottom, it is a very exceptional case that requires the surface sand to be any harder for a light casting than for a heavy one. Making castings by rolling the pattern over in flasks, to form the bottom part of a mould, does not require the mechanical skill or experience required to bed in a pattern, and the manipulations are easier and simpler in getting the surface and bottom part of a mould to right conditions by rolling over and then bedding in; but as circumstances and shop customs more or less control the matter of rolling over a pattern or bedding in, men, to be good green sand moulders, should be just as able to successfully make a good smooth casting by bedding in as by rolling over.*

---

* Valuable notes upon "rolling over" and "bedding in" are given on p. 146, Vol. II.

MOULDING LARGE AND SMALL PULLEYS.

WHEN making pulleys of various sizes, a shop should be supplied with as good patterns and rigging as possible. In making small pulleys, the work he can do in a day depends more upon the convenience of the rigging than upon the man. In moulding large pulleys some firms have full patterns; but for a special size, or when there is only one or a few to make, the sweep and core-box are used to save pattern-making. For very large pulleys with double arms these are necessary.

The cut on page 35, showing sweeps, brick-work, and half section of mould, having two sets of arms formed with dry sand cores, represents different modes of sweeping or moulding pulleys. By the use of this rigging, pulleys from five to twenty feet diameter, and of any width of face required, can be moulded. For forming the outside face, there are two ways shown. One is by using the sweep, *X*, and the other by using a segment, the elevation of which, and a hook for drawing the pattern, are shown at *B*. With this segment the outside can be moulded either in green or dry sand. To mould with dry sand there would be required an iron bottom ring and cheeks, or side flasks. After it is all rammed, hoist the outside off by handles on the bottom ring, or plate. The mould can then be blacked and run into the oven to dry. The sweep can be used either for green or dry sand, as well as for loam. If the mould is swept up with loam, or dry sand, it is better if possible to hoist off, and should the oven not be large enough, it could be drying on the floor while the inside of the pulley is being moulded. *O* shows the stakes driven down alongside of each handle to guide the outside off and on. The four plates, one of which is shown under the lifting-ring, are to insure a good bearing for the outside to rest on, should the sand joint be disturbed by walking on, or otherwise, when sweeping the inside.

When sweeping with green sand, a hole is dug in the floor to about the depth of face required, and a wooden curb, or a piece of boiler iron, is used as a support for ramming the sand against, so as to make it solid. After this the sweep can be worked around to form a true face, which can be made crowning or straight as desired.

When swept up with loam, the outside of the pulley can be made smooth and true, so as to save turning up in the machine shop, if so desired. For very large pulleys this is worthy of consideration.

When moulding the inside of a pulley, the same principle is involved, whether there are one or two sets of arms. The double sets make the moulding more complicated and risky, but in the hands of a good moulder there is little danger.

There are two ways of making arms; one is with dry and the other with green sand cores. The making of the inside will depend upon whether the outside of the pulley is formed with the segment, or with the sweep. Should the segment be used, the inside of the pulley, when the arms are formed in dry sand core as shown, will require to be moulded first, so as to have a bearing for the segment to be rammed against. When the arms are made in dry sand cores, the cores should not be made any larger than is required to give them a body sufficient to be handled with safety.

The cut shows one core resting on the bottom level bed, which is formed with a sweep. There is a projection on the upper side, and also one on the top arm core, so that when both come together they make a hub formed of dry sand cores, and the space between the upper and lower core is filled with green sand. The inside is also rammed and formed with green sand wherever the arm cores do not fill up.

When the face of a pulley is wanted more than three feet wide, the arms would come so far apart as to make one continuous hub, very heavy, unless the center core should have a deep chamber to take out as much weight of iron as possible. Should the hubs be wanted separate, they can be made so by using a flat covering and bottom cores, the same as shown at *H, H,* for forming the bottom and top of the hub shown. In order to let the iron run from the top hub down and into the lower one, there can be risers or flow gates connecting the two as shown at *A.*

The arm core box, *P,* is used for forming the hub, arms, and inside face of the pulley with all green sand. The depth of the box is made the same as the face of pulley wanted, and is spaced off according to the number of arms required.

A double set of arms can be made with the green sand cores, with almost the same surety that a single set can, providing the face of the pulley is not too deep. There could be two sets of cores made, one being on top of the other.

The green sand cores could not be used with safety when the segment is used for forming the outside, nor would it be practicable to attempt to use the double set of cores unless the outside of the mould was made so that it could be hoisted off, and out of the way, as in the plan of the brick-loamed mould shown. Then set the green sand core, using for your guide a mark made on the bed with the core sweep, *W.* When the bottom set of cores are placed on the bed, the upper set of cores can be placed on top of the lower ones without any trouble. In this way it can be seen if there is any crushing, and the joints of the cores can be made up so that there will be no fins on the casting. This core sweep, *W*, is also used for giving form to the green sand that is rammed between the dry sand cores, when they are used to form the arms as shown.

To make the neatest-looking pulley casting, the green sand arm cores are the best when they can be used with safety, for when a casting is made of part green and part dry sand, loam, or cores, each will leave its own trade mark on the casting. The green sand part will swell more or less, according to the pressure of iron when the mould is being cast, but the dried part of the mould will not swell, so when the casting comes out it will have an uneven surface. The different colors of green and dry sand, loam, or cores, on a casting make it look badly; as if it had been made in sections. There are several ways to make a covering for the top of the rim and hub, also arrangements for bolting or weighting down. The first is to have a level dry or green sand cope ; the second, a loam plate ; and the third to make some cores to cover the rim, as shown covering the hub at *H*, and have a cast-iron flat plate to lay inside of the covering cores on top of the sand, or cores, that form the arms and inside of the pulley. This plate is used to lay the weights on to hold down the inside part of the mould when being cast. To hold down the covering cores, small weights are used. Sometimes the rim is cast all open, and the hub and arms, or inside, are the only parts weighted.

The gating or pouring of such castings is generally done (if the center core is large enough) through the center of the core to gates cut into the bottom of the print, so that the iron fills up the mould by coming in at the bottom of the hub ; or by dropping the iron through runners from the top, as shown at *A*. The iron spindle shown, when used for sweeping large pulleys should be held at the top by a brace stretched across the mould, and fastened to two upright timbers sunk into the floor three or four feet away from each side of the mould. If near enough to the side of the building, there could be a swinging arm made to reach out to the spindle, to hold it firm and steady.

The arm pattern in the core-box, *P*, is set into the middle of the outside frame, and after the core is rammed up and ready for the box to be drawn, by hitting the arm at the end *R*, to start it, the pattern can be pulled out easily through the hub end, *D*; after which the outside box can be taken away.

In making these cores, a cast-iron beveled edge plate, the shape of the inside of the box, and made so as to have about ¼″ clearance all around the inside, is set on a level board, or a hard bed of sand. The box is then set on, and the core rammed up nearly to the arm, which is then put into the box and the sand tucked under it even and firm. At this point the moulder must be careful, as in making the cores in this way, the arm cannot be got at to finish it, or to fill up soft places after the core is made. The advantage of making a core in this way is, that when setting in the cores there is no danger of crushing the arms, or of having fins on them, which must be chipped off when the casting comes out, which is likely to be the case when one half is formed at the outside surfaces of the core, *T*, *T*. To lift or hoist these cores, there can be lifting-hooks, or nuts cast in the anchor or lifting-plates, the lifting-hooks being made so as to come up even with the top of the core. When nuts are used, long screws, as shown at *E*, *E*, are used, and when the first core is set in the mould they can be taken out and used for the others.

The cuts 4 and 6 show the plan of making pulleys with a draw-ring pattern. In this way any face required can be moulded from the same pattern.*

At 6 is shown the mode of casting a pulley having a face

---

* Plans for making straight and curved pipes off from draw-pulley patterns are illustrated on pp. 250 and 253, Vol. II.

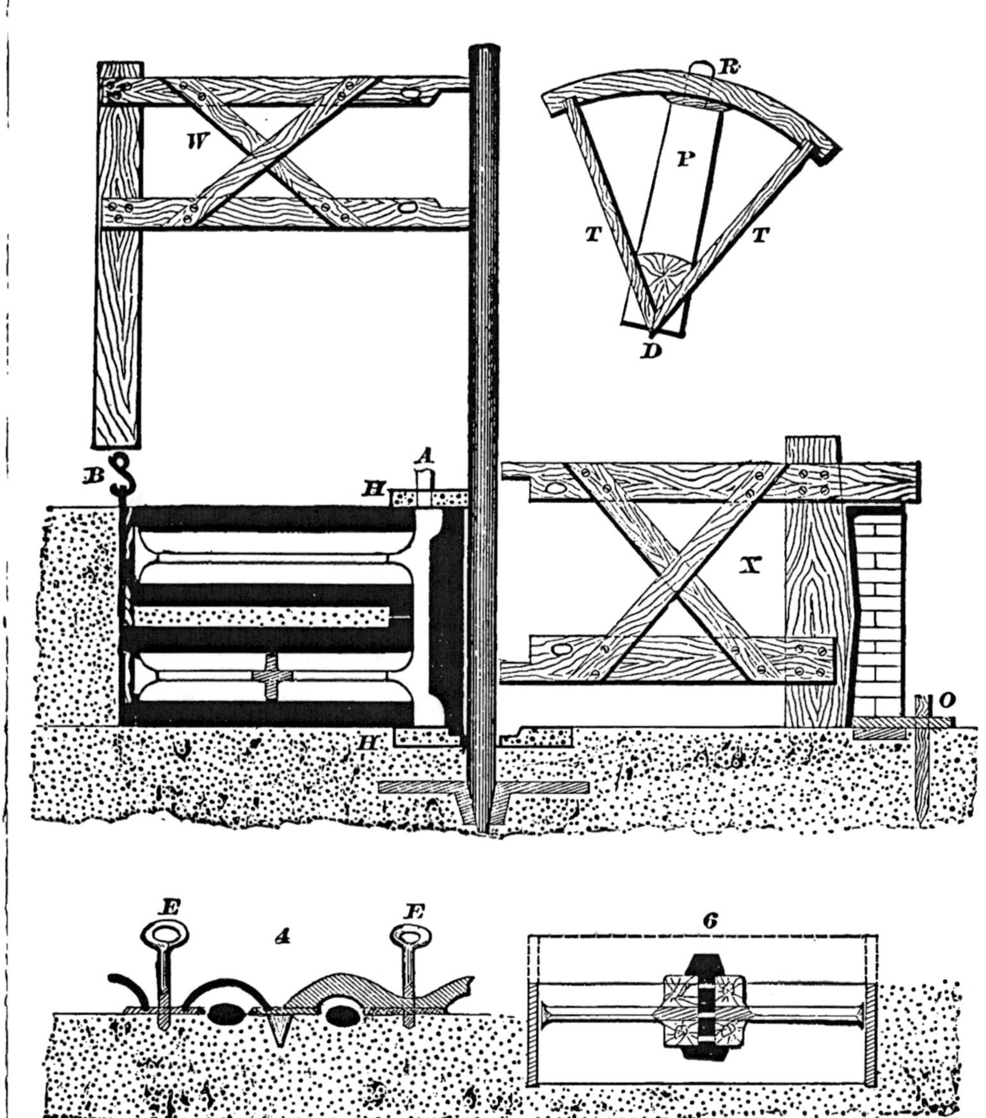

RIG FOR MOULDING PULLEYS.

wider than the pattern. In moulding this a hole is first dug in the floor and the ring pattern set in, leveled and rammed up to about the center of face required. The loose arm and hub are then bedded in. The dotted lines show the distance the ring pattern has to be drawn, in order to have the arms in the center. The pulley can be cast with the rim open, or covered with a cope, as desired. It is best to make the faces about ½″ higher than wanted, so as to give stock for the machinist to true up.

The hub shown is arranged to readily change the core prints to any size wanted. The hubs have a hole drilled through their centers, the same diameter as the holes in the center of the arm pattern, and there are wooden plugs driven into the hubs which project on the side or face that comes next to the arm, and centers the hub. The core prints have also projections turned on them the same diameter as the hole in the hubs, so that a moulder working on pulleys need not be running to a pattern-maker every time he wants to change the size of prints.

At 4 are shown two ways of making the anchor or lifting plates. One style has a wrought bent rod cast in them, reaching from one plate to the other. The second plan is to have a cast-iron rib reach from one to the other—a plan I adopted, and find it to be more reliable and to make a stiffer plate than the wrought-iron rods. The oval, black spots represent the arm between the plates.

When a double set of arms are wanted in smaller pulleys, there are a number of ways in which they can be moulded, but as a general rule, foundries do not rig up to make double arms, there being so few ordered. When one is wanted the rigging is got up with as little labor as possible. In some instances the lower set of arms is made with cores, or a flat core is made inside the ring pattern, having one half of the arms and hub formed in it, and the other half is bedded in green sand. Before the arm pattern is drawn the flat core is set over the arms and staked through holes made in the core between the arms. The core is then taken out and the pattern drawn, after which the arms are finished and the core set back. The pulley is rammed up to where the upper arms are wanted, and the rest of the moulding is the same as in making a pulley with one set of arms. Another way of making the lower set of arms is to have single cores with half the arm and hub formed in them, and when the arm pattern is drawn the single cores are placed back, guided by stakes or sand-marks made by laying the core on top of each arm before the pattern is drawn.

Although using the cores as described is a quick way of forming the lower set of arms, it does not produce as good looking casting as when they are formed by the following plans. In some cases foundries have used a regular anchor plate for the bottom set of arms, and when the castings come out the anchor plate had to be broken in order to get it out of the casting. When there is time to make the rigging, loose plates having nuts for screws, or lifting-hooks cast into them, are used. These plates are set between each arm, and the pulley rammed up 6″ or 7″.

A plate having holes to correspond, so that the screws or hooks can pass up through and be wedged, is bedded on the sand. The pully pattern is then drawn and the flat plate, having all the loose plates wedged up to it, is hoisted out. The arm pattern is drawn and the core lowered back, after which the pully pattern is gently set back. The wedges are now loosened, the flat plate taken out and the upper arm and the rest of the pulley is rammed up and finished.

Another way is to have holes in the upper anchor plates, and by having two sets of arm patterns, ram up the whole pulley. Long bolts with threads cut on each end are used to bolt the lower loose plates to the top lifting-plate, by which the whole core is hoisted out and the lower arm finished. The core is then lowered back and the nuts taken off. The top portion is then hoisted out and the upper arm finished. The bolt holes in the sand are enlarged and the top portion lowered down to its place.

The following dimensions are from what are termed a light, a medium, and a heavy set of pulley ring draw patterns, from 10″ up to 48″ in diameter. The face of these patterns

LIGHT.

Diameter.	Thickness.
10″	$\frac{7}{32}$″
48″	$\frac{14}{32}$″

MEDIUM.

Diameter.	Thickness.
10″	$\frac{9}{32}$″
48″	$\frac{16}{32}$″

HEAVY.

Diameter.	Thickness.
10″	$\frac{11}{32}$″
48″	$\frac{18}{32}$″

generally runs from six to ten inches, and, to draw them, holes are drilled through the pattern within $\frac{5}{8}$″ of the top, and hooks instead of screws are used. In making a set of these patterns they could be swept up in loam, or in green sand, by using a segment attached to an arm having a hole at the radius wanted, to fix on a stake driven into the sand; or the arm could be attached to an iron spindle.

There are some things that a foundryman should think of before starting to make a set of draw patterns. One is, that a poorer grade of iron can be run into heavy pulley castings than into light ones. Should a No. 2 iron, that can be turned in a heavy pulley, be run into a light one, he might be looking for cracked arms, or a blessing from the machinist that tried to turn them. Where competition is sharp it is best to have a light and a heavy set of patterns so that customers can have their choice ; but if you can make them believe that a heavy pulley will wear longer, it will be more money in your pocket. When an establishment intends to make nothing but pulleys, it is better to be fitted up with what are called split pulley patterns, which require a pattern for every width of face wanted. They should also have the best of flasks to make them in, by which means they can be made very fast. But for the jobbing foundry, the draw patterns are the best, as fewer patterns and flasks are needed, and the expense is nothing compared with the cost of getting up a stock of split patterns, and the necessary flasks.

For the proportion of either straight or crooked arms, there can be found full figures given by Chordal in the AMERICAN MACHINIST, July 23, 1881.

As regards contraction or cracking of pulley arms, I will say, to prevent the arms cracking select iron having the least possible contraction.*

---

* For further information upon the cracking of pulleys, etc., see p. 255, this book.

FINISHING GREEN SAND MOULDS.

Go into any machine or jobbing foundry, and notice moulders finishing or patching moulds that have been broken in drawing the pattern, and you will see some one mending a corner, for instance, that has been started or broken, by taking his swab and wetting the part to be patched, and then taking some sand and pressing it on the top of the wetted part. Another moulder, not having so large a piece to mend, will swab the part, and then patch on sand with his trowel; or he may be finishing a cope overhead, when ten chances to one he will be raising his trowel for rubbing sand into the holes, and every time the trowel goes with a bit of sand it is sleeked up against the smooth surface, caused by the pressure and sliding movement of the trowel. Although he will see the sand falling down, as fast almost as he puts it up, he will keep on trying until he thinks something is the matter, then he will tell his helper to get him some nails; that the sand is so rotten and poor it will not hold together.

He will then push up some nails to hold the sand. Nails are a useful article, but some moulders will make a casting without using one, while another, in making the same casting, will use two or three pounds, and, perhaps, if he did not use them his casting would not be good. Some moulders will ram up a mould in such a manner that it will not require half the finishing it would require if rammed up by another.

If a moulder thinks he has more time to finish the mould than he has to ram it up, he will hurry or slight the ramming; or he will do this, perhaps, in order to catch the use of the crane, or to get ahead of some moulder on the gager pile, etc. As there are plenty of nails in the kegs, he will whisper to himself that he will not bother putting a rod in that corner, or be particular in ramming it ; for if the pattern when drawn starts or knocks off the corner, he will have plenty of time to patch and nail it. Or he may ram, rod, and vent the bottom in a creditable manner, and slight the cope, by not tucking the bars good, or ramming up the sand solid, and when the cope is lifted off, and a large lump of sand falls ont, he will think of the nail keg and smile. Should the foreman complain about using so many nails he will tell him that the crane jumped, or that the old wooden flask had ought to have been broken up long ago, and if times are good and men scarce, the foreman, to avoid any words, will walk away, and in a short time he will order the nail kegs to be locked up and carry the key in his pocket.

About the first thing a moulder should do after his cope is lifted off—when the pattern is bedded in the floor— is to lay some boards around on his joint so as to preserve it, as there is nothing that looks so slovenly as to see the joints of moulds all trampled and cut up by kneeling on them when finishing.

In drawing out a pattern, the top edge of the mould is always started more or less, and it is the first part that the moulder should give his attention to, by getting it sleeked or fastened down to it original place. In finishing over the mould, if there are any parts that look started, it is best, if practicable, to tear them off instead of just pressing the sand back—even if there are some nails in it—and rebuild or patch it up, not with sand on the trowel, but by using the hands to press the sand with. By using the hands a moulder can unite and shape the soft loose sand on the solid sand in a shorter time and in a great deal more reliable manner than by patching it on with a trowel. How much more mechanical it looks to see a moulder, when mending a cope overhead, take the sand in his hand instead of on the point of a trowel.

When the parts are made solid take a little wooden straight edge, shape and smooth off the extra sand, and go lightly over with the trowel or finishing tools. In patching on sand a moulder's fingers never caused a cold shut or scabby casting; but too much sleeking with tools often does so. In such castings as thin pipes or plates, it is better to have the fingers go easily over the sleeked or finished mould, and then rough the surface up a little, as iron will lie quietly on a rough surface, when it would boil or bubble against a smooth, sleeked surface.

If any part of a mould to be mended is too dry for the sand to stick to, dampen it by taking a mouthful of water and blowing it out in a fine spray. When water is swabbed on an extra dampness, or mud, is formed, so that when the hot iron is poured into the mould, although it may have surface sand the right temperature to lie on as this surface gets heated, the heat soon reaches this extra dampness or mud, and, as heat, when it comes in contact with dampness is sure to raise steam, and the sand not being of a body strong enough to hold the pressure, it will escape by lifting the sand on top of it, and passing up through the iron will cause it to bubble, or cause the casting to blow.

If the swab is used it should be only on the surface, for then, when the steam is made, it has only to raise the iron to pass up through, and, if there is a scab on it, it will be a very light one.

For heavy or light casting sleeking or swabbing must be done in an intelligent manner, if good castings are expected.

THE DRAWING DOWN OF GREEN SAND COPES.

The surface condition of a green sand cope, while the iron

is being poured into the mould, entirely depends on the mixture and nature of the sand, and the heat it is subjected to. Any section of a cope surface that is exposed to the direct heat of the metal for over twenty seconds, requires the sand to be strong and close and gaggered well, having as little sand under the gaggers as possible, to keep the sand from been drawn down. I noticed in a recent issue of the MACHINIST, the assertion made that, with a plate 2″ or more in thickness, the cope will be baked hard as a brick by the intense heat before the iron reaches it. I only wish that such was the case, for it would save work and anxiety for the result of many large castings. I have made moulds in green sand that were not safe to cover with a green sand cope, and have covered them with a loam plate, fearing that the green sand would draw down. This is caused by the sand exposed to the heat getting dry and dropping down on the rising iron, which, when the casting comes out, shows lumps and sand-holes in the cope part. There are several ways of securing a cope surface, to a great extent, from drawing down. For instance, mix some flour in your facing sand (about one to sixteen or twenty), or wet your sand with clay wash, and, before closing the cope, sprinkle the surface over with molasses water, or beer.

Above everything, keep your risers and feeding heads air tight, so that there is no chance for the air in the mould to escape, except through the venting and the sand. Then the rising metal will compress the air above it sufficiently to keep the sand from being drawn down by the heat, if the mould is not too long in filling up, so that the pressure is released by the air having time to escape through the vent-holes and the sand. There is also such a thing as pouring a casting too fast, so as not to give the air a chance to escape as freely as it should, thereby lifting your riser cover and weights, and letting the air rush out and start your mould blowing. I have made castings where I have nailed the surface of the cope over with nails, keeping the heads about one-eighth of an inch below the face of the sand, and in some cases have had the nails even with the face of the mould, so as to insure the cope against being drawn down. When not feeling sure of this, I have made the facing sand strong with flour, and wet it with clay wash, and when the cope was finished made it very damp with molasses water, building a fire with shavings and chips under the cope, until the surface of the mould was dried like a dry sand mould. When I thought that none of these extra precautions would keep a green sand cope from getting dry or burnt, I would then use a loam or dry sand cope, or covering.

MOULDING BEVEL AND SPUR WHEELS IN GREEN SAND WITHOUT A PATTERN.

GEAR moulding is something that nearly every jobbing and machine foundry has something to do with. Gear wheels are often broken in use, and are readily replaced if the patterns are at hand in some foundry near by. But when to replace them it is necessary, as is often the case, to send a good ways, there is generally great delay before the casting is received.

Some mills and factories keep in stock wheels to replace those that are liable to be broken, which is a very good plan, as the expense of carrying a few wheels is trifling when compared with the loss of having a machine, or sometimes the entire works, shut down until a new wheel is procured.

The cut represents a plan for moulding or sweeping up a bevel wheel, the pattern work for which can be made in a very short time compared with that required to make an entire pattern. The sweeps and segment could be made one day, and the gear cast the next day, unless in the case of a large wheel. The advantages of sweeping up such wheels, where there are only one or two wanted, is the saving of making a full pattern, and the saving in time. Of course it takes more time to mould a gear with the sweeps and segments than where a full pattern is used, but this extra time is not to be compared with the labor required to make the pattern.

In sweeping up gears in this way the spindle seat is first sunk and trued up, and, if the wheel is large in diameter.

it is best to have the top of the spindle held firm, by having a brace attached to it. Then a coke or cinder bed is placed, as shown, after which the sweep *X* is fastened to the spindle. A bed for the hub, *B*, and loose arms, *P*, also for the inside face and the top surface of the teeth, is swept up, and, at the same time, the joint is also swept. The hub and loose arms are now set on. The surface of the bed and joint having been well sprinkled and sleeked up with parting sand, the cope is set on and rammed up. When the cope is lifted off, the sweep *S* is fastened to the spindle, having the edge *H* just bearing on the joint, so that when the sweep is revolved it will not disturb it. If the joint is disturbed it will leave a fin over the tops of the teeth.

The depth of the teeth and rim, also the thickness of the plate or web, as well as the hub core print, are then swept up. The segment *Y*, having an arm screwed on to it, is then secured to the spindle, as shown, and the teeth are rammed up. The tops or joint edges of the teeth are better for having some long slim nails pushed through the sand.

The vents should be carried into the cinder bed, instead of being carried off at the joint, as is generally done when there is a full pattern to mould from. The reason for nailing and venting in this way is that when the cope is lowered down, to see if any of the teeth will crush, if some of them should touch hard, the nails will help to hold them from being broken, or from sticking to the cope when it is hoisted off again. If vented at the top, the fins, of which there will be more or less at the tops of the teeth and at the joint, will be sure to get into the vents; but when carried off through the cinder bed, the joint all around the flask can be rammed so as to prevent any run-outs and burning the flask.

There should not be less than six teeth on the tooth segment. The more teeth, the quicker will the moulder get the teeth rammed up.

S

X

B

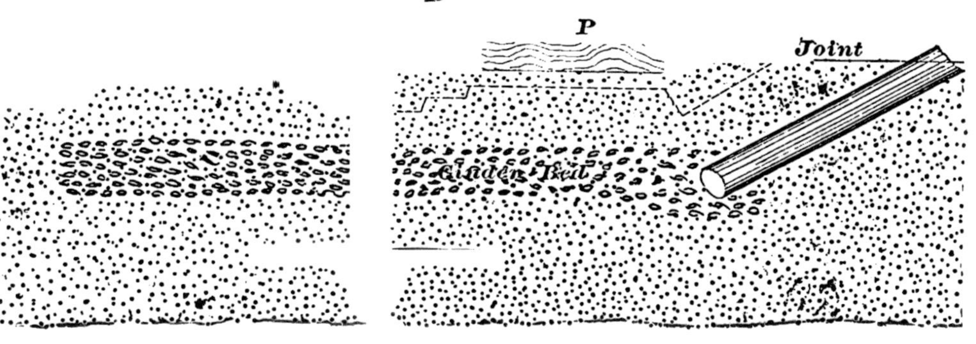

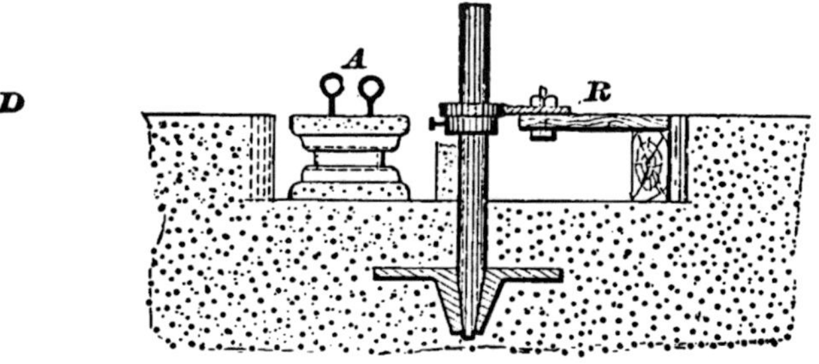

DEVICE FOR MOULDING SPUR AND BEVEL WHEELS.

Great exactness is required in sweeping up gear wheels after this plan, in order to get the right number of teeth, and also to have the last tooth rammed up of the same width and space as the others. After the bed for the segment to lie on is swept out, it is best to go around with the segment, marking the ends of the teeth on the bed, so they may be counted, and make sure of having the last tooth come right, before starting to ram up the teeth. *W* and *D* show the last tooth, which is liable to come larger or smaller, when the diameter is not set exactly right. When marking off on the bed before ramming up the teeth, should the last tooth be found to leave too large a space, as shown at *D*, the diameter must be made less. Should the space be too small, as shown at *W*, the diameter must be increased.

The amount that the radius is changed to bring the last tooth right, is approximately one-sixth of the measurement that the tooth is too large or small. Should the space be $\frac{6}{8}''$ (or $\frac{3}{4}''$) too large, the radius should be made $\frac{1}{8}''$ smaller, after which it is best to go around again and see if the change has made it right.

The cut *R* shows the process of sweeping up a spur wheel. A level bed is first made, and then the tooth segment fastened to the arm, which is of different shape than the one shown for sweeping up the bevel wheel. The collar, which is for supporting and allowing the arm to revolve around on it, is held firmly by the set screw shown. After the teeth are all rammed up and finished, the dry or green sand cores can be set on the level bed to form whatever style of arms are wanted. The arms are shown at *A*. The spindle is then taken out, the hole filled with sand, and the center core set in. The mould is then ready for the cope to be set on.

The cope should be rammed up on a level mould board, or on a level bed of sand. After the cope has been tried on and off it is then set back, or closed on for the last time. The weights are then hoisted on, and the bars wedged down if necessary. The pouring basin, or runner, and feeding heads are made, after which the mould is ready to be poured.

Referring to the spacing of the teeth in such a way as to come out correctly—that is, to have the last tooth and space of the same dimensions as all the others—care must be taken when ramming up and changing the segment. The best plan is to depend upon the marks made at the ends of the last tooth on each side; then, since by ignoring the marks, guiding altogether by setting the end segment tooth up against the face of the last tooth moulded, there is danger of having a thin or thick tooth at the conclusion, after these marks are correctly made, it is best to cover them over with pieces of board, paper, or anything to prevent them from obliteration. Another plan sometimes adopted as a guide for changing the segment, is to shake out flour on the sand bed, so that when the segment is lifted a perfect impression of the teeth is shown, and by carefully keeping loose sand from the bed there will be correct impressions on the bed by which to reset the segment.

IMPROVEMENT IN MOULDING GEAR WHEELS, PULLEYS, ETC.

PROBABLY in no branch of the iron business has so little been done to assist, by mechanical appliances, the skill of the workman as in that of moulding. In the main, the moulder goes about his work to-day substantially as he did thirty years ago, his success depending on his skill in the use of the simple tools then known to the trade, rather than to the advantages of new appliances.*

As showing, however, that some thought has been expended in the direction of improved methods of moulding, we illustrate herewith a patent device of R. B. Swift, of Cleveland, Ohio, a practical moulder of long experience, for moulding such work as gear wheels, pulleys, and similar pieces, from a sectional pattern, which we are informed has been adopted by some large manufacturing concerns to their satisfaction, not only in the saving of time, but in the quality of the work produced.

In drawing the segmental pattern used in making a casting there is always the danger of tearing up the mould. Further, it is sometimes very desirable to make a casting in green sand of such form that it would be impossible to draw the segment directly. The object of this device is not only to provide against the breaking down of a mould with a plain pattern, but to provide for using sectional patterns of such forms as cannot be directly drawn, such as crown-faced pulleys, grooved friction wheels, etc.

Referring to the engraving, which represents a mould in

* The author is happy to state that for the last two years rapid progress in this line is being made.

process of being made, the use of this device may be explained. The segment pattern is shown attached, and, as will be noticed, the spindle is embraced by two half boxes, which are made to accuratelv fit it. These boxes are fitted to be moved in jaws by means of the screws *E, E.* In use, the segment pattern—whatever it may be—is screwed in place, as shown, so as to sweep *approximately* the proper radius. The spindle being in position, the radius is corrected, so as to be *exactly* right, by means of the two adjust-

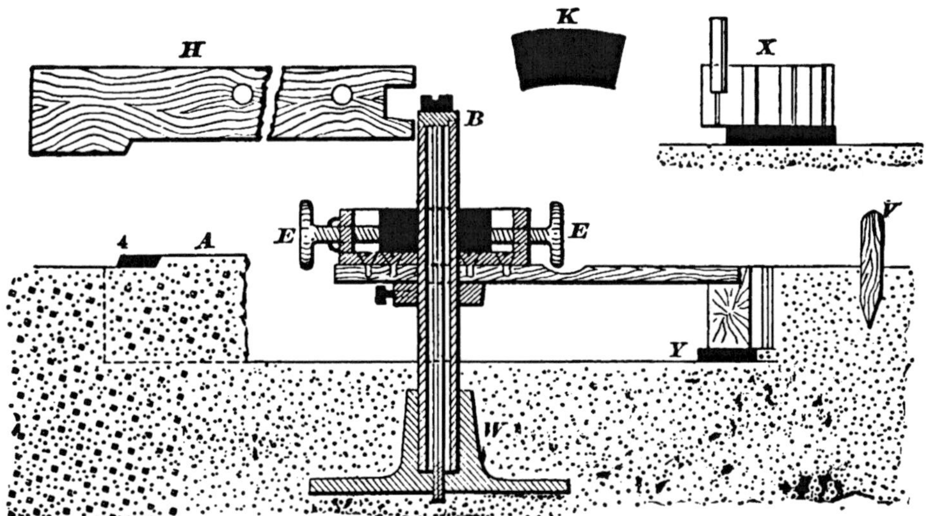

SWIFT'S PATENT DEVICE FOR MOULDING GEAR WHEELS, ETC.

ing screws referred to, the manner of doing this being so apparent as to need no explanation. The advantage of this almost instantaneous means of adjustment will commend itself to any one accustomed to doing this class of work.

This feature alone would seem sufficient to demonstrate the value of the device; but perhaps the most valuable feature is its adaptation to the following purposes: Let it be desired to make a casting which has some projecting parts that would render it impossible to draw the segment straight

In this case either of the screws may be turned back, drawing the pattern either towards or away from the center, as may be desired, until it is free to be drawn without the possible danger of breaking down the mould. When used in this way the opposite screw to the one being used remains stationary, and serves as an accurate stop or guide by which to quickly reset the segment.*

The methods or riggings for moulding, illustrated in this article, are those in use for moulding gear wheels and pulleys in the shops of the Cuyahoga Works, Cleveland, Ohio, and which I am kindly allowed to present by their permission. The upper cut shows the process of moulding a spur gear wheel having a top and bottom shrouding on it, using only a small segment and arm core box to form or make the casting from.

The difference between moulding a wheel having a shrouding and one that has none, will be better understood by referring to the article entitled "Moulding Bevel and Spur Gear Wheels in Green Sand, Without a Pattern," page 45. In moulding this shrouded wheel, the tooth segment X requires to have one tooth loose and long enough to come down on the sand bed, so that every time the segment is drawn it can be replaced exactly, according to the marks made on the sand, as described in the article referred to. The flat loose segment *K*, after the level bed is made and spaced off, is set on and the tooth segment is set on top of it as shown at *Y*. The teeth are then rammed up and the tooth segment drawn, after which the segment shroud is drawn in, and then replaced and the tooth segment reset. Then more teeth are rammed, and so on until the circle is completed.

To form the top shrouding on the wheel, the sweep *H* is secured to the spindle and a solid hard bed is swept up, as

---

* Other devices for machine moulding of gear wheels are illustrated on pp. 242, 262, Vol. II.

shown at *A*. The outside edge, 4, could be formed by having wooden segment pieces laid all around, if so desired, although this is seldom done. After this bed is finished, the cope is set on and rammed up, and being well staked, is then lifted off. A hole the right diameter and depth is then dug out, the bottom bed swept up, and the teeth and lower shrouding formed as described. After this, green or dry sand cores are set in to form the arms and hub, and the cope is closed on, having the stakes as shown at *V* for a guide.

Mr. J. F. Holloway, the president of the Cuyahoga Works,* has designed a spindle for such class of sweeping, that will not shake or turn over from the weight of a heavy sweep. The spindle is made of a heavy tube from 2″ to 4″ diameter. The outside is trued up and the ends faced off. The spindle seat, *W*, is bored out straight so as to be a good fit, and the hollow spindle is set into it. At the bottom of the spindle seat there is a $\frac{5}{8}$ or $\frac{3}{4}$ hole bored, and a thread tapped in it, and when the spindle is set in, a turned washer, having a projection set down into the spindle, is placed as shown at *B*. Then the long bolt, having a head on one end and a thread on the other, is set in and screwed down tight. By this means a spindle from six to eight feet long can he held as firmly as if there was an arm or brace attached to the top of the spindle, as is generally done with spindles that have heavy sweeps attached to them.

The next improvement, and one that is worthy of note, is Mr. Swift's (a foreman of the works) rigging for moulding double-armed pulleys, entirely of green sand. The cut represents the lower arm as being moulded. The draw rim pattern has been drawn and the core hoisted up. The arm and hub pattern is then drawn, and the core lowered down into place. The draw pattern is then set back, and the iron or wooden wedges, Nos. 2, 3, 4, 5, 6, 7, are withdrawn. The top lifting plate is then hoisted off, the loose anchor plates

* The Cuyahoga Works sold out Jan. 1, 1887, and is now known as the Cleveland Ship Building Co. Mr. Holloway died Sept. 1, 1896, leaving behind him very many admirers of his ability and ennobling

remaining down, after which the upper arm and balance of the pulley is rammed up in the ordinary way. Instead of hoisting these cores with a chain, as shown, there is a three-winged cross used.

The cut *D* shows a plan of the lifting plate having one of the loose anchor plates wedged up to it. I think that the advantage of this rigging can be seen without further description.

The next and lower cut shows the manner of moulding some large pulleys, which were cast in segments and bolted together. When these pulleys were completed, they were found so true that they were only ground on the face by using a stone suspended by a rope.

In moulding these segments there was a full pattern used, but the outside face of the castings was formed by using an iron casing, well filled with vent holes. On the casing there was about 1″ thickness of loam swept, the process being shown at *M*. While this is being dried in the oven, the inside, arms and hub, of the casting is moulded in green sand, as shown, the top of the pulley being kept about even with the level of the floor. After the lower half of the segment is rammed up and the joint at the center of the arm made, the iron cope is set on, rammed up and staked. The screws or bolts that are for holding the arm and face pattern together are then loosened, and the face pattern drawn back. After this the cope is hoisted off and the arm pattern drawn, and this part of the mould finished. The cope is then lowered down to place, the face pattern set back, and the balance of the mould rammed up. The face pattern is again drawn back, and this inside part of the mould finished, when the casing is taken out of the oven, lowered down, as shown at *P*, and, after it is pushed up against joints *O* and *T*, sand is rammed up at the back of it to the level of the floor. Covering cores, *S*, are then set on, and the mould weighted down ready for pouring.

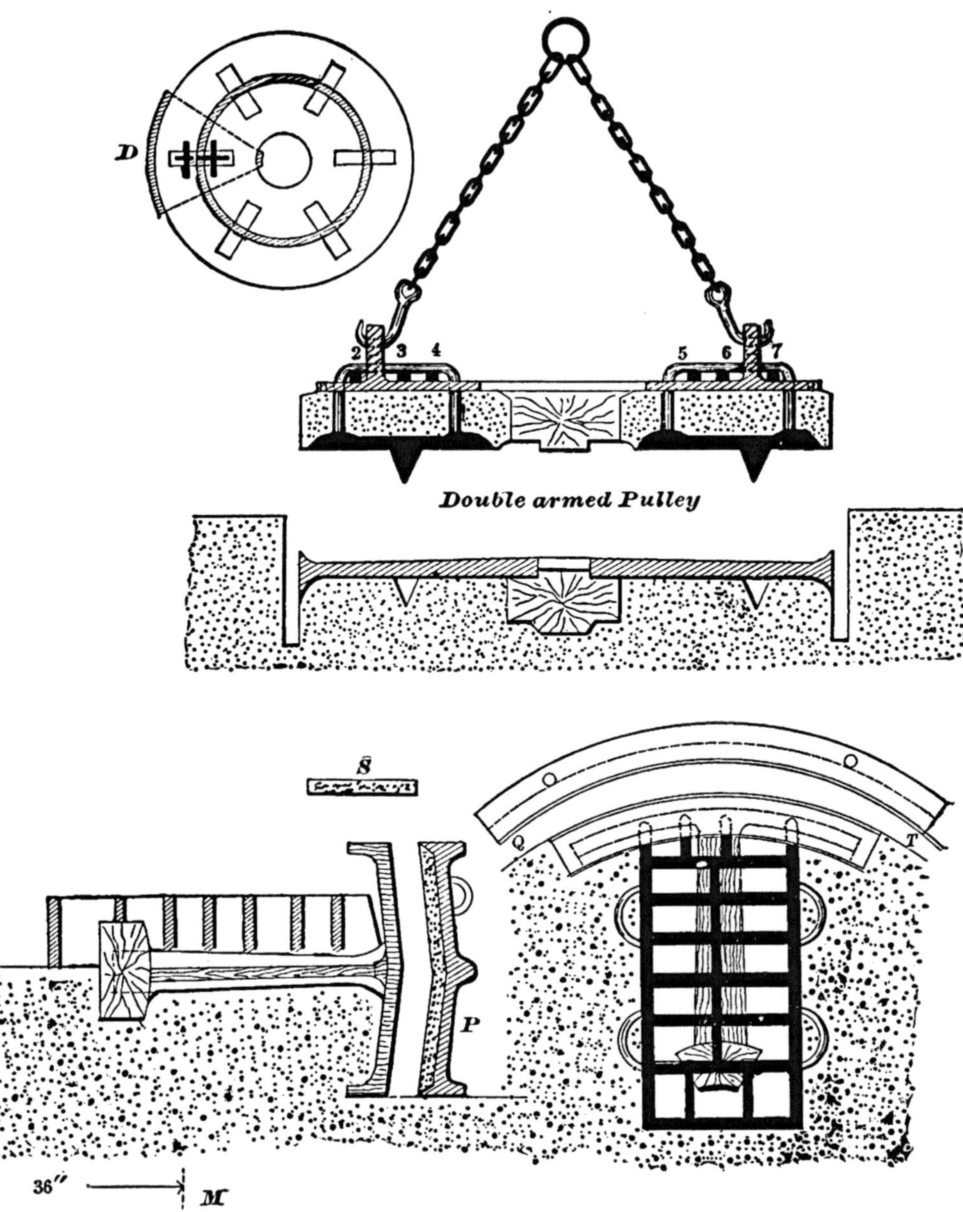

DEVICE FOR MOULDING PULLEYS.

VENTING GREEN SAND MOULDS.

VENTING a mould with a vent wire is done to allow the free escape of air and gases in the sand, together with the steam generated by the liquid iron coming into contact with damp sand. New sand will not stand ramming as hard, and needs more venting than old sand, by reason of the additional life and gases in it. Sand mixed with sea coal or minerals needs still more venting because of the increased gases. Were it not to provide for the escape of these gases, air, and steam, moulds could be rammed as hard as iron, and have no blowing or scabs. The bottom of moulds often requires the most venting, because it is the part which takes the longest time to be covered with a body of iron, and when covered, is surrounded mostly by the iron.

Plain copes are vented more to allow for the easy escape of the air confined in the mould, than for the escape of gases or steam in the cope sand. Plain coped work, poured with hot iron, requires less venting than if it were poured with dull iron, as the hot iron has life enough to force the air up through the pores of the sand. If the iron were dull, the compressed air at some spots, not finding as ready relief, would hold back the iron, and by the time the pressure (or the air) escaped, the iron would be frozen, so that when the casting came out, it would show smooth, flat hollows in the cope part.

There is very little difference in venting plain copes for heavy or light casting, as regards the closeness of the vents. Light work should be vented to the surface of the mould, so as to allow the air to escape rapidly, while heavy work that requires a pressure of air to keep the cope from being drawn down, should be vented one to two inches from the surface. Copes having any pockets, flanges, or projections in them, require such places to be well vented. Moulds poured very fast require the same treatment.

I shall never forget an incident that occurred in a shop where I was working some sixteen years ago. It was before I had made up my mind to study cause and effect in foundry work. A moulder, doing some of the best work in the shop, was making a plain cylinder flatwise; it was about two feet in diameter by three feet long. He had a full split pattern with which to mould the outside; while for the inside the core was made on a wooden core barrel, full of small vent holes, with nails driven into it to hold the sand. The core barrel had iron trunnions on the ends, which rested on iron horses, extending out so as to hold the sweeping board. The sand was packed with the hands in the barrel, and the sweep made the required diameter. The first two or three castings that he made were lost, on account of top portions of the core lifting up off the barrel. The man did everything he could think of to save them, using longer nails, making his sand tougher, and using very thick claywash. Above everything, he gave great attention to having the vent fired while pouring. In making the next casting, through some excitement, the vent was not fired until the mould was full. This casting, to the astonishment of us all, was a good casting. The cause of the previous trouble had been in firing the vent before the mould was full, causing an explosion which started the core.

There are moulds that require a bed of cinders under them, and it is as essential to know at what time the vents should be fired as it is to know how to vent the mould. Take the case of a mould having projections, green sand core, or any portion that the iron does not cover for some time when first going into the mould, it is here sometimes best not to have the vent pipes lighted until the mould is full of iron, and for such classes of moulds pipes of a large diameter are better, and at least two pipes should be connected with a cinder bed, since by so doing the danger of the vent exploding is avoided. Two outlets will also cause a freer circulation of air, and in so dangerous a class of moulds, the vent pipe should be located where there will be no danger of flying sparks of iron entering them while the mould is being poured. When the vent explodes before all the bottom surface of a mould is covered with iron, the pressure of air and foul gas created finds relief with a sudden force, and presses itself into all openings and available space, so if all the bottom surface of the mould be not covered, such explosion may drive the air and foul gases through the vent holes, and be likely to lift or start any portion of the mould that may not be covered with iron; the result of this would give a scabbed casting, or would start a mould to blow. Whenever vent pipes are lighted for an ordinary line of casting, they should be lighted at the top of the pipes, for by so doing the current of explosive gases is drawn from under the mould to the pure atmosphere, where they can escape and burn freely; and if these gases cannot be drawn to the top by burning shavings at one side of the top of the pipes, it is best not to fire at the bottom of the pipes, but wait until the mould is full of iron. A good supply of vent wires of all sizes is needed by every foundry, since their liberal use has saved many a casting that would otherwise have been lost through hard ramming, or wet or inferior sand.*

---

* A valuable chapter which should be read in connection with this is found on p. 155, Vol. II.

MOULDING KETTLES WITH A DRY SAND COPE AND GREEN SAND BOTTOM.

ORDINARY kettles are usually made in loam, having the bottom cast up.* The engraving shows a plan of casting the bottom down, which will make a sounder kettle, that will last longer than one cast with the bottom up. The size of this kettle was about six feet diameter and three feet deep. The outside was swept in the floor with green sand, and the inside was made in dry sand, swept up on the carriage and dried. The cope was made in two sections, and bolted together as shown. The reason for doing this was, that the

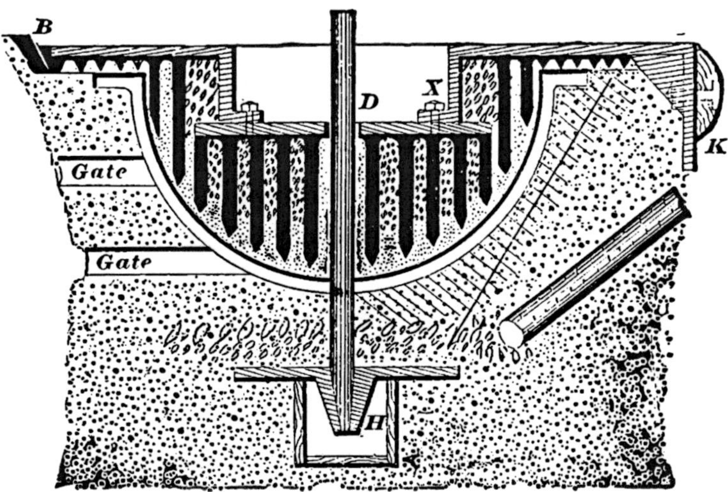

prickers were too long to drive and make a good plate ; also, the ring *X*, formed of two pully patterns, made a stiffer plate than one cast flat.

---

* Moulding kettles in loam is described on p. 149, this book.

In getting up this rigging there are two improvements I made, and found them to be of value. The first was the mode of turning the cope over; and the second, a plan for closing the cope down true on the bottom.

Instead of sweeping a face or seat on the bottom and a corresponding one on the cope, to fit into it, as is usually done for such work, and which is shown at *B*, I had a hole cast in the center of the plate, one-quarter of an inch larger than the size of spindle, and leaving the spindle in its seat *H*, the cope was lowered down over it, and when within an inch or so of being down to its place, we saw that the space between the joints was alike all around. Just before the two joints touched each other, we saw that the spindle was in the center, as shown at *D* and *P*. In doing the job this way, if the spindle is in the center of the hole when the cope is swept up, you can rely on the casting having an equal thickness all around. After the cope is lowered to its place, drive down some stakes at the four handles, take out the spindle, hoist off the cope, and fill up the spindle hole with green sand. Then lower down the cope the second time using the stakes for a guide.

This plan saves work in sweeping out seats or guide faces, which usually takes a deal of time, and when done are not reliable, especially in large loam work, as the expansion of the plates when heated in the oven will crack and displace the brick-work more or less, causing the seat to be out of true.

In sweeping up the cope, coke and cinders were put in around the prickers, so as to leave about ten inches of sand on top of it. At the joint where the short prickers are, fine cinders were used.

The dry sand used for sweeping up the cope was made very open, as close sand will not make so smooth an inside. As the casting was only 1″ thick, after the form was roughly swept, some gaggers were driven into it, so as to hold the face of the mould from dropping, should it get jarred or cracked when being rolled over.

The rigging generally used for rolling over such copes as this is shown at 2, 3, and 4. The trunnions 2 and 3 should be cast below the level of the plate, to balance the weight of the sand and plate, and when turning throw a rope over the

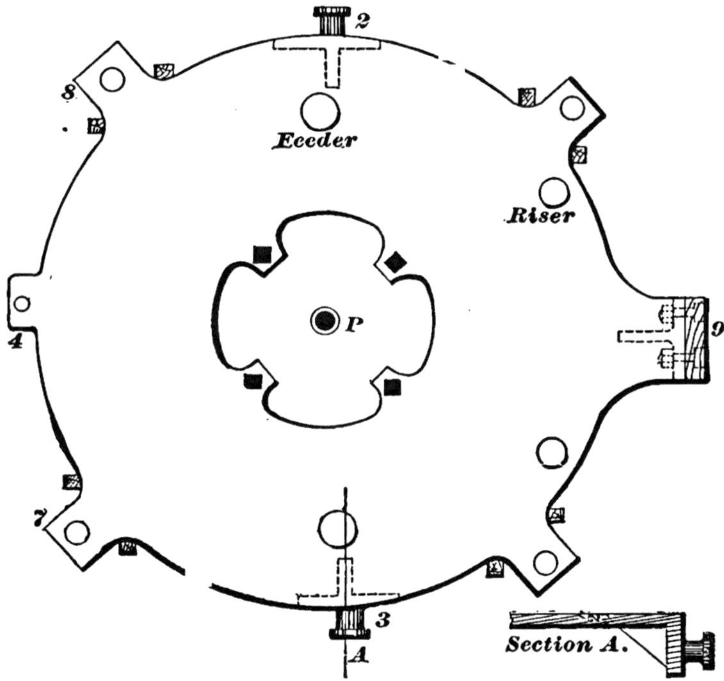

lifting beam, and hitch on the handle 4. In this way the plate can be let go over easier than if left to turn on the trunnions alone.

When there is a heavy body of sand, or when the plate is large in diameter, the following plan is the best: Hitch the chains into the handles, 7 and 8, and let the foot, *K*, which has a wooden roller bolted to it, rest on a strong plate or block of wood. Then, as the crane is hoisted, the roller will cause the plate to turn over with ease and steadiness. Should there be any fear of a jump when the plate is on the balance, pnt some blocks under the foot *K* to catch it ; also have some men with long sticks to reach the top handles to steady it over.

Why I give the preference to this plan for turning over plates is, that the plate is resting on three bearings, which will spring it less than when it is turned over with two bearings, as when rolled over on two trunnions.

In sweeping or moulding the bottom in the floor, a coke bed was laid under to carry the vents, and the sides were swept up first, a space being kept open around the bottom for the moulder to stand, and for the sand to fall into. When the sides were finished the bottom part was swept up, and the casting gated as shown. For a pattern two sweeps were made, one for the bottom and one for the cope. The cope plate was cast 2″ thick, with plenty of vent holes in it.

DROPPING OF GREEN SAND COPES.

THE expression "dropped," or "fell," made use of in a foundry, will turn every moulder's eyes in the direction indicated. Even if he is drawing a fine tooth gear wheel, ten chances to one he will give a squint to see who is the victim, if it knocks down every tooth to do it. Why it is that this is the case, is only known to moulders. There is nothing that will cause the countenance of a moulder to change, and that will make him look as if he had lost his last friend, so quickly, as to have all, or a portion of the cope of a mould which he has been working on for a day, or perhaps two or three days, drop out when he is closing his mould.

If the cope is closed by hand, this may be caused by not lifting it level and steadily; or if hoisted with a crane, the chain may jump. One or the other of these is about the only excuse a moulder can make for such an accident.

The foolish manner in which some moulders will gagger copes will cause them to drop quicker than if they had never put a gagger in them.

Not long ago, an old moulder was ramming up a cope that had fallen out with him, and going to see what was the matter, I asked him what made it drop. His only answer was, "It fell out." I told him, by the looks of things, there was no question about that part of it, and seeing by the manner in which he was gaggering up his cope, he did not know what the trouble was, I asked him if he knew what he set gaggers in a cope for? He answered, "To hold the sand up." Taking a gagger and setting it in, I asked him which he thought was the heaviest, the iron gagger or a piece of sand of the same dimensions? Well, he thought the gagger was.

I then asked him what cross bars were put in a cope for, at which point I saw he was getting indifferent, and told him I was speaking for his benefit, and still insisted on the question. He answered, "To hold in the sand." Having still hold of the gagger with my hand (for if I let go it would fall down), I asked him to give me the longest gagger he had in his pile, which, when set in, did not come up 2″ between the bars that were cut out; and seeing the questions had answered my purpose, I walked away from a man who was evidently wondering why he did not think of these simple questions before.

Copes dropping from just such causes are every-day occurrenees with moulders who have worked a life-time at the trade.

In ramming up copes that have the bars cut out so as to require gaggers, or should there be a body of sand to be lifted with gaggers, the moulder should remember that iron gaggers are heavier than sand, and if he wants to lift a body of sand with them, the gaggers should be long enough to have two-thirds of their length up between bars, as it is the sand rammed between the bars that holds the gaggers, and it is the gaggers that lifts the hanging sand below the face of the bars, in some cases. When there is over two inches of sand to be lifted, there often should be, to assist the gaggers, some wooden sticks, or "soldiers," as they are usually called —a name that must have been derived from the resemblance of the sticks, when in position against the bars, to a company or regiment of men in line.

I once came near getting struck by a green German helper because I told him to go to the pattern shop and get some "wooden soldiers." He looked at me, and wanted to know if I thought he was a fool, and to make matters worse, several men working near were laughing at him. Before I could get the soldiers, I had to go for them myself.

In using soldiers, they should not have too large a surface on the end that comes next to the pattern, for when this is the case, the sand is liable to drop, or be drawn down from them and cause the casting to blow.

When using soldiers for making heavy castings, the gaggers should be set first, and then a good inch of sand put over them before setting the soldiers. When set in this way they can be used larger, which will make them of more service. If there is a heavy body of sand to be lifted, the soldiers can be nailed to the cross bars. Wooden soldiers will lift a heavy body of sand better than iron gaggers, which can be proved by trying to pull one of each out.

When soldiers are used over the surface of light castings, their end surface should not be over $\frac{3}{4}''$ square, and they should have a good $\frac{1}{2}''$ of sand under them. The space between wooden or iron bars has a great deal to do with the amount of hanging sand a cope will lift.

Copes that are made for jobbing castings should not have the bars over six inches apart, and the bars should be at least 7″ deep, so that they will stand to be cut out and still leave width enough to be gaggered.

Copes that are made for special patterns, if the castings are light, can have nails driven into the chamfered edge of the bars if necessary, doing away with the use of gaggers or soldiers for lifting or carrying the sand. For plain, ordinary light castings, if the bars are clay-washed and not over $\frac{3}{4}''$ from the face of the pattern, there is little danger, if the sand is in good condition and rammed as it should be.

If the sand is burned much, and the moulder is not allowed to put in new sand enough to renew it, he will have to gagger it more, and should select the thinnest and lightest of gaggers. He will also have to ram his mould harder, to keep the sand from dropping out of the cope.

There are moulders working on good work who will make casting after casting without a hole caused by the cope dropping after it is closed, while others cannot make over two or three castings without trouble of this sort, for which they always have an excuse.

The way some moulders ram up a flask will be cause enough for all their trouble. They will have some 8″ of sand in the cope for the first ramming, making no difference when there are flanges, pockets, or anything else to ram over or around, giving every piece the same treatment with a heavy rammer. For the second ramming they will have only 3″ or 4″ of sand to ram through, over which they will spend as much time as they did ramming the first course of 7″ or 8″. For a finish, they will go over the top in a loose, careless manner, and then vent it. This may be a quick way of ramming up a cope, but it is far from being a reliable way.

In ramming plain copes, from 4″ to 5″ of sand is plenty for the first ramming, and which should be even and solid. For the second ramming you can put in 7″ or 8″, and go over it in half the time. Then, with a butt rammer, make the top solid, for it is the butting that will make the sand compact between the bars, so as to hold the gaggers or soldiers in a firm manner. In this way you can depend on having a good lift, and the sand will stay where it belongs.

When there are pockets, flanges, or projections to be rammed over, a light hand rammer should be used so as to ram in and around them evenly, and not get the sand so hard as to cause blowing, but still solid enough to hold the sand from dropping. Should the pattern be so constructed as to require a deep cope, the same treatment and precaution should be used so far as dropping is concerned.*

* An article which should be read in connection with this is found on p. 155, Vol. II.

MOULDING KETTLES IN GREEN SAND WITHOUT A PATTERN.

DIFFERENT styles of kettles require an entire change in the manner of moulding them. In some foundries, where they have a standing order for a special-shaped kettle, they have good patterns and other arrangements for making or sweeping them up in loam. In making kettles of almost any form, there has to be more rigging-up and expense incurred than in making ordinary castings. Loam jobbing castings are worth more than green sand ones, on account of the extra labor, time, and fuel; and in a great many instances they are made in loam simply to save the cost of patterns. One reason why so many kettles are swept up in loam is the expense of a full set of patterns. Notwithstanding this, however, a set of patterns is sometimes made for a single casting.

The engraving represents a plan employed in making a kettle which was wanted in a hurry. Instead of sweeping it up in loam, it was swept up in green sand. The only rigging needed was the lifting frame *X*, which was made of pulley rings and a few pieces of wood for the bars *H*, *H*. The holes, Nos. 2, 3, 4, and 5, are for bolting the frame up to the cope, as shown at *T*. The size of this kettle was 7 feet 2 inches at the top, and about 6 feet at the bottom, the thickness of which was 2″. The sides tapered in thickness from 1½″ up to 1″, the depth of the kettle being 24″. The casting was poured with two gates, 4½″ wide and 1¼″ thick, as shown at *W*.

The cuts *P* and *O*, show the wooden sweeps laid against the spindle for sweeping the outside and inside of the casting.

In starting to mould a casting like this, it is necessary to have a coke or cinder bed under the mould. At *B* is shown a seat for holding the spindle. When the spindle is in place and the sweep *P* fastened to it, sweep up the shape of the inside as shown at *D*, and, instead of shaking on wet parting sand to make a joint, it is best to fasten paper on the side with small nails. Before setting the lifting frame, there should be three thin flat plates, about 6″ square, set on the bottom to keep the frame from sinking into the face of the mould.

After the frame is rammed up, set in the four bolts. These bolts are better to have a nut on each end, which makes them more solid than having a hook on the lower end; as when the weight of the core comes on the hook it is liable to yield and crack the core. The top of the bolt above the nut should be squared for a wrench, to hold them from turning around while screwing up the nut.

At *M* is shown one of the four bars of iron resting on the top of the frame and wedged under the wrought iron bar. This makes it certain that the melted iron will not raise up the core, as is often the case. As it is, should the core rise, it would have to lift up all the holding-down rigging.

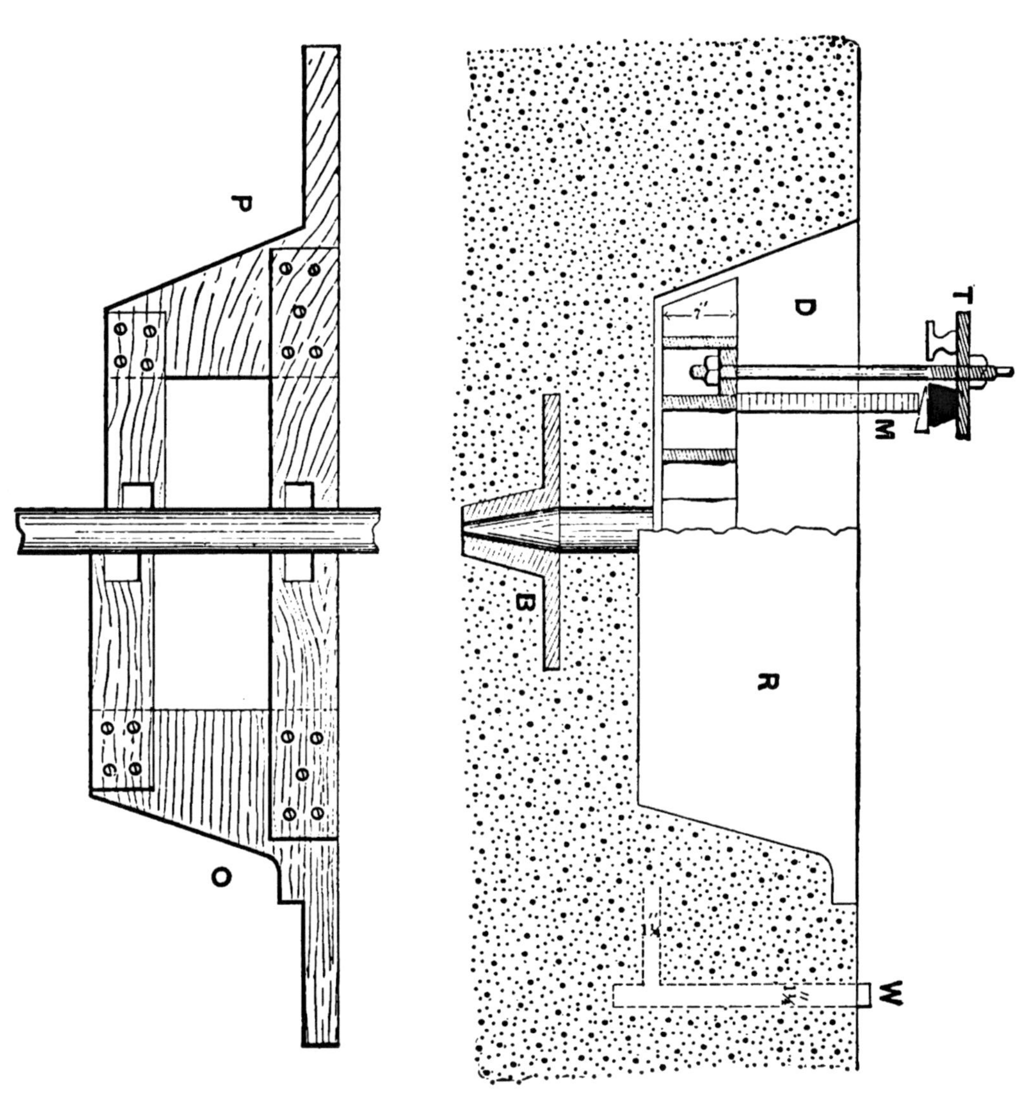
P
O
D
T
M
B
R
W

When the core or inside is rammed up level with the joint or top of the kettle, then set on the cope, and after it is staked, rammed up, and vented, lay on four rails or bars for bolting the core up to the cope, as shown at *T*. Two of these rails are placed side by side, so as to carry the weight of two bolts. These bars must have their bearing on the sides of the flask, and be raised up high enough to clear the wooden cross bars.

Before screwing down the nuts, place some heavy weights on the rails, say as much weight as will bend down the bars equal to what the weight of the core would bend them; and, while the weights are on, screw down the nuts solid, after which the weights can be taken off. Then wedge between the upright bar *M*, and the lifting bar, and also all the wooden cross bars. Bolt up any heavy core or body of sand this way, and you can depend on there being no cracks or openings in it.

After the cope is lifted off, set back the spindle, fasten on the sweep *O*, and dig out about 4″ of sand all around the sides and bottom. This will leave room enough for packing sand, and sweeping it out the shape of the outside of the kettle, as shown at *R*. Kettles 12 feet or more in diameter can be made in green sand after this plan, provided the building and crane are strong enough to lift the cope. The casting will be as sound and solid as one made in loam—if anything, better; as in moulding kettles of this style in loam, the bottom is usually cast up, while this one is cast down, which will always make a sounder bottom.

MOULDING ELBOW AND BRANCH PIPES WITHOUT A PATTERN.

THERE are often cases where a party wants a special piece of pipe in a hurry, for which he is unable to find a pattern. Almost any shaped pipe can be made, at very little expense for pattern-making, by a little extra work in the foundry. Let a clear sketch of the pipe as wanted be given to the pattern-maker, from which he will make a plate pattern 1½ inches wider than the outside diameter of pipe, the extra width forming a bearing for the sweeps, *X*, which are to form the core and thickness of pipe. This pattern should have pieces nailed on where the flanges are wanted, as shown at *B*, *B*, and an extension of five or six inches beyond all flanges for a core print. From this pattern cast two open sand plates. These cannot be counted as involving extra expense, since they would have to be made if the core was rammed in a regular core box. When all is ready, ram the core sand a little larger than the size of core wanted, and take the smallest sweep and strike off the core, following the shape of the plate when possible, and when not, as at *D*, use the trowel. When the core has been gone roughly over, sprinkle it with water, and sift on core sand, using a fine sieve. After this has been packed evenly by the hands, it should be gone over evenly and steadily with the sweep and the core slicked and finished.

The sweeps should be made $\frac{1}{16}$ inch to $\frac{1}{8}$ inch larger than the size required, as the slicking will make the cores smaller.

In the bottom half of core put some nails, letting them stand out as far as possible, and have the thickness sweep clear them. These nails will help to hold the thickness on when the core is turned over. The top half of core will not need any. After the core is dried, wherever metal is wanted, rub on loam, or other mixture, that will sweep well and bake hard, and strike off with sweep to size. Have as many flanges as there are to be on the pipe, turned up and cut in halves; and, to make sure of setting their faces true, it is better to have a strip of wood fastened to them, as shown at *A*, and when setting the flanges (before sweeping the thickness) drive a few nails through this strip and into the core.

The loam swept on for thickness of metal will require drying, unless the core is very hot. In using this core as a substitute for a pattern, dig a hole in the floor (that is, if you cannot get a flask to suit), bed the bottom core, set on the top half, and rub down the prints, or put sand between them, till they caliper round. The top half of the core should have hooks, so as to lift it up with the cope. When the cores are drawn knock off the thickness, paste the halves together, and let them dry while finishing the mould.

If the pipe is to be one inch or more in thickness, another plan would be to saw out a number of half-circle pieces, as shown at *H*, place them over the cores, 4 inches or 5 inches apart, and when rammed and the cores drawn, cut out the sand between them. This would save sweeping the thickness on the core, and in some cases might be the cheapest and the best plan.

Ordinary size pipe, when time cannot be spared to cast plates, may be swept up on a wooden plate made the shape wanted, and having the flanges nailed to it to keep them in place. Sometimes, when three or four

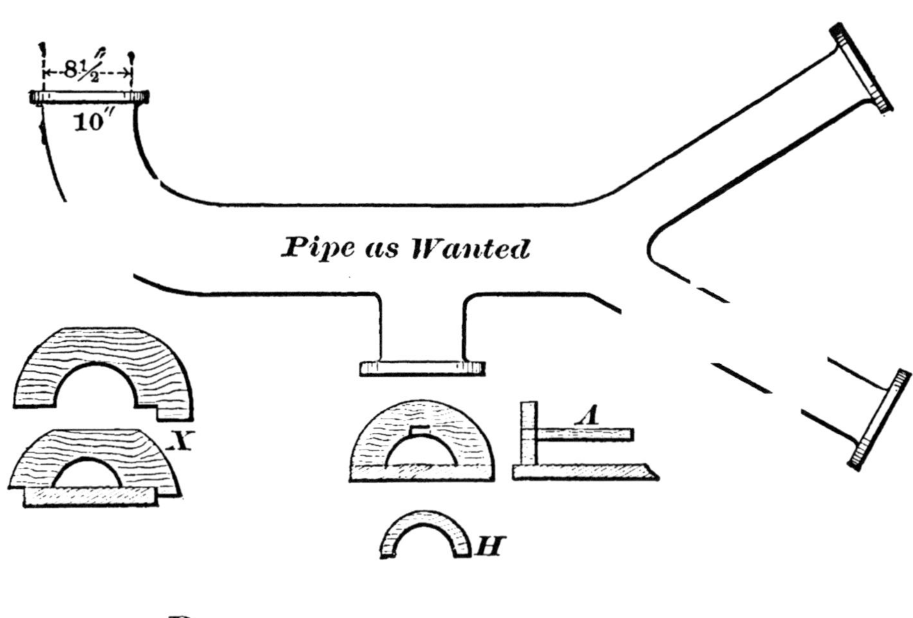

Core Plate D

castings are wanted and it is not undesirable to make a wooden pattern it is best to make a core expressly for use as a pattern not sweeping on a thickness, and when the castings are all made, the core can be broken up and the sand used again.

RAMMING UP THE TEETH OF GEAR WHEELS IN GREEN SAND.

THERE is, perhaps, nothing requires more careful and even ramming than the teeth of gear wheels. The sides of almost any casting can be swelled without attracting attention, but if the face or sides of teeth are swelled, it will appear at once. In some shops great attention is paid to haying each tooth the right size, and in some instances every tooth is tried with calipers to detect swelling. Moulders will sometimes make the teeth of wheels exactly the size of the pattern, while others will be from $\frac{1}{32}''$ up to $\frac{1}{16}''$ larger than the pattern. Teeth can be larger than the pattern and yet show no signs of strains or swelling, even if the moulder has been very particular in ramming, for if he did not ram solid, being afraid, perhaps, the teeth would not draw well, would be scabbed if rammed solid, or perhaps from his established practice of light ramming, it is sure to occur. It may sound odd, but there are few moulders that ram alike. One will ram heavier or lighter than another, and their castings apparently show no difference, but if they are tested with calipers and straight-edges, or weighed, then it is easy to see who rams the hardest. It sometimes is a good thing to be accustomed to ram hard. But for general jobbing work the moulder accustomed to ramming lightly will have the fewest bad castings.

Some moulders can ram a mould light or heavy as they

choose, but they are few. In ramming up small spur teeth from 2″ pitch down, the hands only should be used; some moulders will *press* the sand in, while others throw it in, and raise an inch or two every time. It would be a hard matter to decide which is the best plan, since the moulder who pressed the sand in, if told to throw it in, would probably make the teeth swelled, because he is not used to making small-teethed wheels in this way. In making very small-teethed wheels, where there is difficulty of getting the sand to stand, it is a good thing to use some new dried moulding sand, and after it is screened very fine, dampen it with some beer or water, and then, with a wooden or iron roller, roll it back and forward over the sand, until it is well mixed; this will make the sand tough and give a good body to it.

Teeth from 2″ pitch up are generally made by throwing in 2″ or 3″ of sand, and then, after the outside is rammed, use a rod or small pin rammer to ram in between the teeth. The larger the pitch the more solid should the ramming be made; the bottom, or first course of ramming, should be rammed the most solid. When ramming sand between teeth there should not be over 3″ for a ramming, and it should be rammed even, and as firm as the sand will allow. Before throwing in sand for another course of ramming, the loose sand should be all scraped away, and any soft sand pressed down by using the fingers; this will help to avoid soft spots between the course of ramming. Teeth that are rammed solid should be well vented; facing sand should be used stronger for the root and sides than for the face of the teeth. A plan that works well in making nice-looking teeth, is to use as strong a facing sand as the wheel will stand between the teeth, and then for the face of the teeth use nothing but fine-screened common heap sand; this common sand on the face of the teeth will allow the pouring of the wheel with duller iron, and still retain a sharp face or corners. When the wheel is cleaned, the inside of the teeth will peel, and by rubbing coke or a piece of grindstone over the teeth's face, it will result in fine-looking teeth. About the most difficult class of wheels to make are bevel gear wheels, since, when bedded in, there is no chance to ram the teeth, as may be done when ramming spur wheels ; sometimes in large bevel wheels having small pitch it is possible, after a bed is made, to lift out the pattern and turn it over, so as to bring the face up, then fill and press the teeth full of sand, then by handling the pattern gently roll it over on to its bed, and pound it down. This plan works very well when the teeth are small enough to hold in the sand when the pattern is turned over, but for patterns that cannot be thus managed, the moulder will have to ram up the teeth by having a bed made the shape of the bevel of the pattern, from 2″ to 4″ below the face of the teeth, this space affording a good opportunity to ram, and enabling the moulder to get his hand underneath. Ramming up of teeth has to be done more by the sense of feeling than seeing ; in this the moulder must rely on his own mechanical ability as to the amount of hardness that teeth will stand in being rammed. The harder that sand will stand ramming, without danger of scabbing or not drawing, coupled with even ramming and good venting, the better-shaped teeth will be produced on a casting.

CASTING LARGE PIPES IN GREEN SAND.

The plan here described and shown for moulding large pipes or work of a similar nature, involves small expense, being in that respect far cheaper than loam moulds, especially when the oven is not large enough to dry the moulds, and they have to be dried on the floor. The foundry that made these pipes had an order for about a dozen, and they were wanted in a hurry. The pipes weighed about 5,000 pounds each. Their diameter was nine feet, and their height five feet, and the thickness of metal one inch. There was a flange at the top and bottom by which to bolt them together. Elbow pipes that went with the plain ones were cast in loam, and, having worked on both jobs, I will give a description of how these were made when I get to loam work. In making the plain pipes there was a sheet-iron curb sunk into the floor to prevent straining, and to save work in digging and ramming. The draw pattern was made of wood, and was 18″ deep, with four strong draw irons on it. In starting to mould it, a cast iron ring, *X X*, is set level, from which a level bed is made. This ring is never disturbed, so that the leveling by straight edges every time a casting is made is avoided. When the bed is finished, the draw pattern is set down, and cores having the bottom flange formed in them, as shown at *A*, *A*, are set around the pattern. There are two of these cores that the runner cores are attached to, as shown at *B*. These were set a quarter of a circle apart, and when all the cores were set, any open joints were packed with hemp, so as to keep dust or dirt from get-

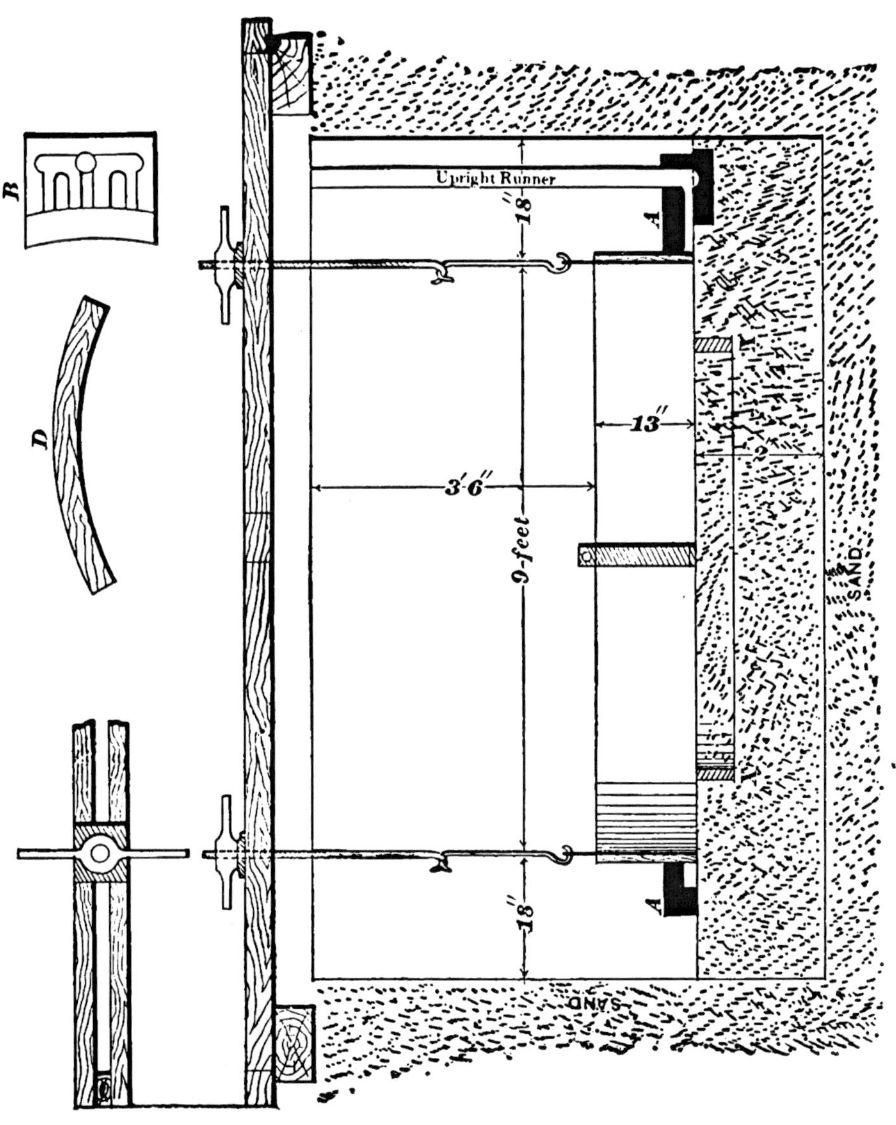
Upright Runner
18″
A
13″
3′6″
9-feet
SAND
A
18″
SAND
B
D

ting to the flange. The facing and sand was shoveled in and rammed solid up to the level of the pattern, and then well vented. When all was ready, a man was placed at each screw-handle. The cut only shows two screws and one beam; but as there are four draw irons needed, it takes four screws and two beams. At the word "Around!" each man turns his handle around once, doing this at every command. In this way the pattern is drawn even. This is a splendid rigging for drawing gears or anything that needs to be drawn level and steady. This pattern was drawn about five inches at a time, until it was raised up level with the top of the curb, which made it the height required. The pattern was leveled at every drawing, and the vents carried up to the top by venting at every raising. After the cope was rammed up and taken off, the segment *D* was bedded all around the pattern to form the top flange. The upright runners were rammed up in the green sand, and had a core placed at the bottom to prevent any cutting. The ramming was light towards the top, so that the iron would lie quiet, and to prevent any straining of the bottom portion, it was rammed more solid.*

---

* The principle here set forth for using draw-pulley patterns for moulding pipes is further illustrated on p. 253, Vol. II.

MAKING AND VENTING BEDS.

THERE are two classes of beds—one is the open sand bed, and the other is a covered bed. When a bed is covered with a cope, it can be made harder than a bed without a cope, since, when iron is poured on to a covered bed, the air in the mould is more or less compressed, and this pressure causes the retention of the gases, steam, and air, and forces them to find relief by passing downwards into the sand below the face of the mould; and the harder this underlying sand is rammed, the more pressure will be required to drive the gases and steam downwards. When the sand below the face of the bed is rammed too hard, so that the gases cannot be forced downwards, they will, when the pressure becomes strong enough, pass up through the surface sand of the bed and through the liquid iron in the mould as an air-bubble passes up through water. Whenever gases, air, or steam have to pass through the surface of a bed in order to find relief, there will result scabbed castings. The vent wire is used to make a proper channel for the escape of the gases, air, and steam. The vent wire in some cases, when not used understandingly, is more hurtful than beneficial. For example, if a moulder venting a bed directly from the face surface of a mould, and, to keep the iron from getting into the vents, only rubs the palm of his hand over them, the holes seem to be all stopped up; but when the castings come out of the sand, the core boys appear, picking up cast-iron vent wires to make core rods of, and a scabbed casting is the result, caused by the hot iron bursting through the thinly-covered vent holes, and causing them to be all filled with iron; thus, instead of the vent holes carrying off the vent from the surface of the mould, they not only create more gas at the surface of the mould, but also in the interior, or deeper portion of the sand. Beds thus vented would be more likely to produce a good, smooth-skinned casting if they never had a vent wire used on them. A covered bed generally requires to be vented, since this class of beds, if made as soft as open sand beds, will sink from the pressure or strain of the iron, when the mould became full, upon the bottom bed, and thus make the casting a deal thicker than it should be. A good example, showing the results of having too soft a bed, is the case of making thin fire fronts, that had a large semicircular flue-cleaning door-hole, two firing door-holes, and two ash-pit door-holes in them. Around the outside edge of these fronts there was a heavy ornamental border, and around all of the door-holes there were chipping strips and also lugs for hanging the doors on. The thickness of these fronts varied from three-eighths of an inch to three-quarters of an inch, the outside measurements varying from three feet by five feet to nine feet by thirteen feet. When these fronts were fitted up, there was much complaint because the fronts were swelled all over the surface and crooked, occupying a machinist from five to fifteen hours longer to fit up a front than would have resulted if the fronts had been made right. Since there were so many lugs, core prints, and chipping pieces on the fronts, the moulders seemed to think the only way to make a bed to mould them on, was as follows: Straight edges would be leveled up, and a bed the full size of the front made.* This bed would consist of all loose, soft sand, running from 6″ up to 12″ deep, and leveled off. On the top of this soft bed the pattern would be placed, and then, with a sledge-

* How to level up straight edges for making level beds is described on p. 153, Vol. II.

hammer and a block of wood, the lugs, border, chipping pieces, and the whole thickness of the pattern would be knocked down into the soft bed. It was a quick way of bedding in the pattern, but when the casting came out it was a botched job—a disgrace to the moulders that made it. The face, in may places, would be covered with scabs, and the casting would be from one-eighth of an inch up to half an inch thicker in some places than others, and by no means straight. This style of sledge-hammering down a pattern, and the making a bed, is one that should very seldom be adopted, as it causes the bed of a mould to be the reverse of what it should be. Since the surface of a mould should be the softest, in order to have the iron lie quietly against it, and to prevent any strains or swell's, the under portion of the sand should be firmly rammed. In the above case, the hardest rammed sand formed the surface of the mould, and the soft sand was underneath. To properly make a bed for this class of work, so as to prevent any straining and swelling, and have a casting as it should be, the following plan can be relied upon : After the straight edges are leveled, dig out below the level of the straight edges about 5″ of sand, and then, with a butt rammer, go all over the surface ; after which fill up with good riddled or mixed sand, till even with the top of the straight edges, then butt-ram this down also. This will make a solid bed of sand within about one and a half inches of the top of the straight edges. Before going any higher with sand, take $\frac{1}{4}$″ diameter vent wire, and vent the bed all over, after which, with the flat of the hands, close up the tops of the vent holes, and then fill up and level the sand with the top of the straight edges; after which, if facing sand is to be used upon the face of the mould, or common heap sand, it is the proper time to distribute it over the bed; then, with some pieces of wood or iron, three-eighths of an inch in thickness, laid upon the leveled straight edges, level this heap or facing smooth and even. There will now be a level bed of sand $\frac{3}{8}''$ above the top of the straight edges; then, after removing the $\frac{3}{8}$ inch thickness pieces from under the parallel or strike straight edge, rap down this raised sand, so as to be even with the top of the straight edges. This will complete the making of the bed; the fire front pattern is now set on the bed, and the impression of the lugs made, the outside corners of the pattern being staked, and the pattern is then lifted off, and holes, about two inches deeper and wider than the lugs, are dug out of the bed, and all filled up again with soft sand. The cope surface of the pattern is shown by chalk marks over all the lugs, core prints, and chipping pieces, and the pattern is now set back upon the bed, and all the projections pressed or hammered down, the chalk marks being a guide to show what portion of the pattern requires knocking, the bed having only been made the size of the pattern inside of border, which runs all around the outside edge. This is then tucked up with facing sand, the joint is now rammed up and made, and, after the cope is rammed, a gutter is then dug on the two longest sides of the mould, about 4″ below the level of the joint, and a long $\frac{3}{8}''$ vent wire is then used to vent under the pattern, so as to connect, and thus bring the gases from the smaller vertical vent holes to the outside of the mould or gutter. Should there be any danger of a run-out, or the sand choking up the gutter vents, there can be some cinders placed in the gutter, and then filled over with sand, and before going to cast, a few holes can be dug down to the cinders, so as to allow the vent to escape.* The difference in making a bed for such a thin casting as these fronts, and a bed for thicker or heavier castings, is very little. The heavier the casting, the more strain there is upon the bed when it is

* If many castings are to be made from the same pattern, a coke or cinder bed is effective in saving the labor of using the long $\frac{3}{8}''$ vent wire.

being poured, hence the harder should be the foundation or the body of sand within one inch of the surface of the bed; but the surface sand should not be made much harder for a heavy casting than for a light one. It is when the first inch of running iron is being raised upon the mould's surface that the scabbing, caused by too hard surface ramming, is generally done.

Whenever it is necessary that the surface of the bed should be harder, then pounded down three-eighths of an inch of sand. The best plan is not to make the surface first higher, and then ram down, in order to have it stand an increased strain, but to ram up firmly within $\frac{3}{4}''$ of the top of the straight edges, instead of the $1\frac{1}{2}''$, as above stated; and, after this solid bed is well vented, and the surface holes closed and the facing sand put on· then use the $\frac{3}{8}''$ thickness pieces to level off with; this would make $1\frac{1}{8}''$ thickness of soft sand, occupying three-quarters of an inch when pounded down. Many moulders make beds for a heavy casting by ramming up within about $\frac{3}{4}''$ of the top, and then putting on facing sand, and afterwards either pound it down with a parallel strike or a butt rammer. They will then vent directly from the top surface,* and to keep the iron from getting into the vent holes, poke their longest finger into every vent hole; after which they go all over the bed and fill the depressions. If these practical moulders would only think a moment, they could not help seeing that the object which they are trying to accomplish is really being destroyed, for, instead of the vents being brought near to the top surface, the action of the finger has closed them up, so that it would be safe to say that there were no vents within $1\frac{1}{4}''$ of the top surface. Another objection to this plan is the irregular resistance of the bed, since it is almost impossible to make a bed of uniform hardness

* The size of vent wire generally used being $\frac{1}{4}''$. For making beds reliable with surface venting, see p. 89, this book.

when one spot is rammed with a strike or rammer, and another with a man's fingers. It takes but little reflection to see which of the two plans is the best. A bed with its vents an even distance from the surface, and the bed itself of an even hardness all over its surface, two important points that cannot be denied as elements most important in order to make a good reliable bed. Beds for open sand castings are, as a general thing, never vented, because there is very little strain upon the bed, and therefore the lower sand can be left unrammed, which leaves the sand so porous that the gases can freely escape downwards. When a casting over 2″ thick requires an exact level and smooth face upon it, it is sometimes best to make the open sand bed having the lower sand rammed and the bed well vented, similar to the bed described for making the fire fronts. One of the most popular classes of castings made in open sand are furnace plates for rolling-mills and blast-furnaces. When a foundry has a quantity of this class of work to do, it is a good thing to use some sharp sand mixed in with the moulding sand. This will permit the beds to be rammed lightly, without using any vent wire upon then. Some places make the surface of the bed altogether of the sharp sand, and, to make the sand peel from the castings, they use water-lime cement dusted on and sleeked, like blacking. The lime is valuable to peel a casting, but the objection to its use on nice work is the dead color which it gives to the skin of castings. The sharp sand used can be either bank or lake sand. The thinner the casting, the softer should the open sand beds be made. For plates about ½″ thick, dig up the bed about six inches deep, and leave it soft; then open it as it leaves the shovel. If the sand is lumpy, it should be riddled. Level this sand with the top of the straight edges, then, if the sharp sand is used for the face or surface of the mould, spread it over, and level it a quarter of an inch higher than the top of the straight edges. Should the casting be from one to two inches in thickness, the sand should be leveled three-eighths of an inch, after which the whole surface should be pounded level with the top of the straight edges. When a bed is struck off, as is sometimes required in order to make a very smooth surface after being pounded down, it is best to use a straight edge having a face of about one inch, and the striking off should never be done by pulling the straight edge along the sand, but it should be done by working it in an irregular manner across the bed, taking pains that every move is a forward one, for, if the edge is allowed to go backwards, it will leave marks upon the bed. The objection to striking off the bed smooth is that it is apt to start the surface sand if not done right. One great trouble with castings made in open sand is, that they are generally very rough or scabby, at the places where the iron was poured into the bed. To prevent this, the bed in front of the basin should be made harder by using the flat surface of a piece of board. When there is a large amount of iron to be poured, the portion made harder should have facing sand for its surface ; and this hardened portion, whether of facing or common sand, must be treated with the same process of venting as in the above cases for making hard beds.

The most difficult class of open sand casting that a moulder makes beds for, so as to prevent the iron from boiling or bubbling, is large loam plates, that require long prickers cast on them. Sometimes it is necessary to have prickers three feet long, and in making a bed for these, it is impossible to use soft sand entirely, since the strain at the bottom portion of the deep prickers is such that the entire upper portion of the bed would be lifted up, so that the intended prickered plate would become a rough, solid mass of iron. The only safe way to make a bed of this class is to lightly ram it at the bottom portion, within about 8″ of the top, and make the remainder of the bed similar to one for a light, thin, common plate casting. But when the long prickers are as near together as five inches, the plate will bubble more or less in any event when being poured, for the large amount of gas formed below the surface of the bed will cause this on account of the depth of the prickers. This trouble could be remedied somewhat by making a coke or cinder bed below the bed proper, and venting the bed into it with a small sized vent wire.

A good illustration, showing the effect and amount of gas that is formed under open sand plates, was when a number of plain plates were wanted in a great hurry, and, since there was not much shop-room to mould them in, they were made very close together, having only about 3″ of space between each plate. The size of plates was 6 × 4 feet, and there was a bed about 6′ 6″ wide and 25′ 6″ long made to mould six plates upon. They were poured, commencing at one end. The first and second plates were successfully poured, and the crane ladle was then refilled, and the third and fourth plates poured with some bubbling and blowing; but when the fifth and sixth plates were poured, they boiled so hard that the iron was thrown three or four inches, and, as it became molten, so that the gas could not escape, its force raised the middle portion of the plate nearly 2″ from the face of the mould. Had they been vented with a large vent wire straight down between each plate, all this trouble would have been avoided, as the gas could thus have escaped, and not been drawn from the plates already poured into the one that was being poured. It might be asked why the bed was not vented underneath. The bed in this case was all right, and soft enough to stand, and required no venting, as the pouring of the first two plates proved; but the trouble was closeness of the moulds. And thus, by not taking into account a simple, practical truth, a result, which should have proved a success, became a lamentable failure.

On p. 85 objections are raised to the practice many moulders have when making "hard beds," of venting direct from the surface with about ¼″ vent wires, and stopping up the surface vent holes with their finger ends. Now, the author does not wish it understood that good, reliable beds cannot be made by "direct surface venting;" but, like everything else in moulding, there is generally a wrong and a right way. In venting direct from the top surface of a bed, such a system should be followed as will not require the surface finger-poking objected to on p. 85. The following is a plan for surface venting which, if followed, will require no finger-poking of vent holes, and should insure beds being reliably vented for work or sand thought to "easily scab." The plan consists simply in ramming solid within about ¾″ of the top of the straight-edge with common sand, and then filling the surface over with facing sand so as to project about 1¼″ above the straight-edge. After this is butted even and lightly down (*which should then leave the facing sand projecting above the straight-edge*), vent the bed closely with a ⅛″ vent wire. If there is a cinder bed below, the vent wire will of course be pushed down to it; otherwise the ⅛″ vertical vents will require to be carried off by close under horizontal vent holes of ¼″ or ⅜″. After the bed is vertically vented, it is then "struck off." The action of the "striking off," and the after smoothing off of the bed with a little fine-sieved sand and a hard-wood smoothing block (which consists of a piece of hard wood ¾″ thick by about 3″ wide and 8″ long), will close the top of the vent holes (as they are small) sufficient to prevent the iron from entering them, and give an even, smooth surface to the castings. In some cases, where, for instance, there is because of an extra hard bed being required or close sand having to be used, extra danger of scabs being produced, the reliability of a bed could be increased by first venting the bed with a ¼″ vent wire before the facing sand, as above described, was put on; then after it was rammed down, let the bed be vented from its top surface with the ⅛″ vent wire, and "struck off" as described above. To prevent the facing sand from getting into the ¼″ vent holes, simply close their tops by rubbing over them with the flat of the hand.

METHOD OF MAKING A HEAVY GREEN SAND CASTING.

NEARLY two-thirds of all castings lost are lost on account of improper methods of making and placing gates and runners. The best method of gating green sand moulds is to make the gates as long as practicable, as it does not take a very long time for iron, when running directly against, or on the sand, to cut or wash it away. It is a good thing, where you have a large quantity of iron to run through a gate, to place cores against your pattern where the wash of the iron is; or mix some flour in the facing sand for the dangerous sections, and, if you can get at it, put in some nails, having the heads even with the faces of the moulds.

The gate and runners, and mode of moulding the casting, as shown in the cut, can be relied on as presenting a safe plan for any casting of a similar construction. This casting weighed about fourteen tons, and, being made in green sand, the utmost thought and caution were required to make a success, as the casting was not a plain block, but a casting that had all the worst features of a green sand mould to contend with, including corners, pockets, projections, and flanges in both cope and bottom part, with a depth of over four feet in the ground.

The runners and gates were all made in cores, and rammed up with the mould. Being obliged to run this casting direct from the air furnace, as there were no large ladles, nor chance to build a reservoir to hold the iron, I contrived the runners and basins, as shown, to let the iron into the

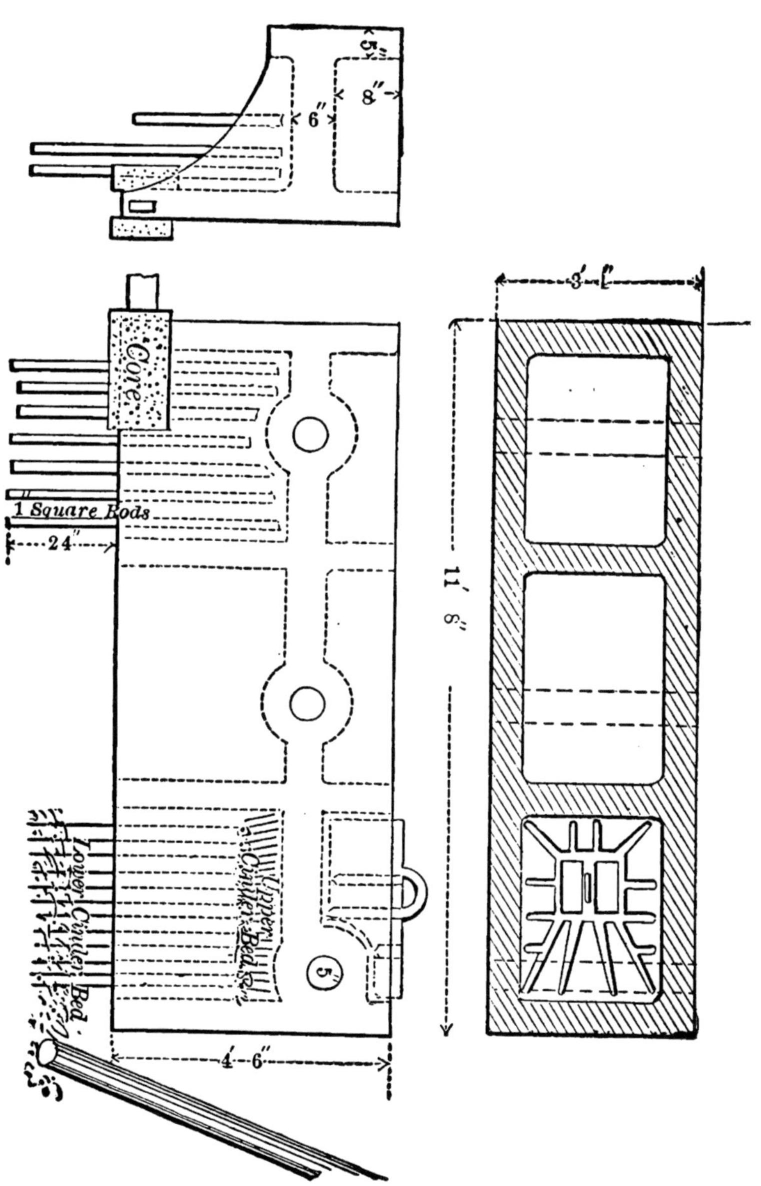
5"
8"
6"
Core
1" Square Rods
24"
11' 8"
3' 1"
Lower Cinder Bed
Upper Cinder Bed
5"
4' 6"

mould, should it come faster than calculated for. At the same time, if any lump should happen to choke up the tapping hole and make it come slower, there would be no chance for the air to blow out at that gate and start the mould blowing. As the iron came down the long spout, it filled up basin 1, which has a core sunk in it to hold back the dirt, so that it is all clean iron that goes into the mould. The basin 2 is the main runner, as it takes the iron to the bottom of the mould. Should the iron come too fast, it will flow over into basin 3. When the basin is full, lift out the iron plug, and the iron goes into the mould. Should the iron come slowly afterwards, so as not to flow over into basin 3, and the mould is not filled up to the level of this gate, the air cannot escape, as there will be eight inches of iron in lower angle gates.

Where iron went into the mould, there were set cores, the shape of the pattern, to prevent any washing. The cut shows how the projections were rodded and vented. All corners were well nailed, and rodded with small-sized rods, also vented with a fine vent wire directly from the face of the mould, which is a good plan to adopt in any green sand corners, that are liable to scab, and the sand for the face of projections was mixed, one-third sharp sand, and rammed lightly. Instead of having wooden bars to lift the pockets out of the cope, iron frames were made the shape of each pocket, and bolted to the cope. The facing sand for the cope was mixed one to twenty of flour, and when the cope was finished it was well wet with molasses and water, while a fire of shavings and chips was made under it, until the surface was dried like a dry-sand mould.

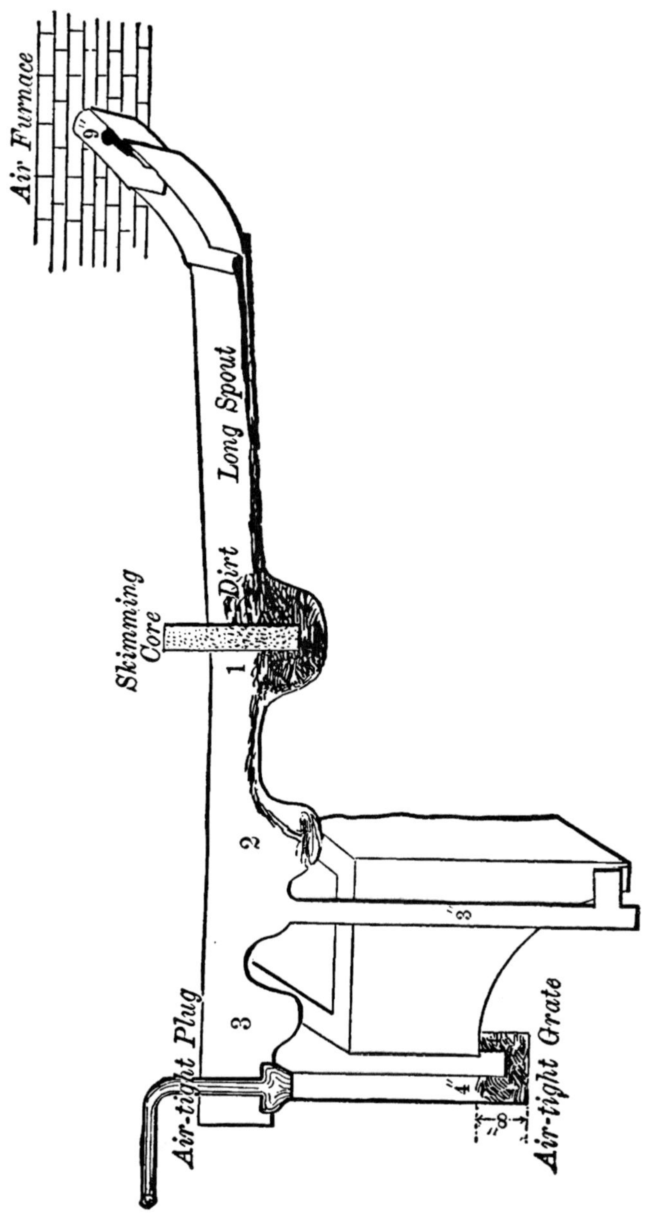
Air Furnace
9"
Long Spout
Dirt
Skimming Core
1
2
3"
3
Air-tight Plug
4"
8"
Air-tight Grate

IRON AND WOODEN FLASKS.

For a foundry to have a good supply of flasks is one thing, but to keep them in repair is quite another thing. In a jobbing shop, especially, it is a serious trouble and expense to keep the flasks in order and mated. A moulder is

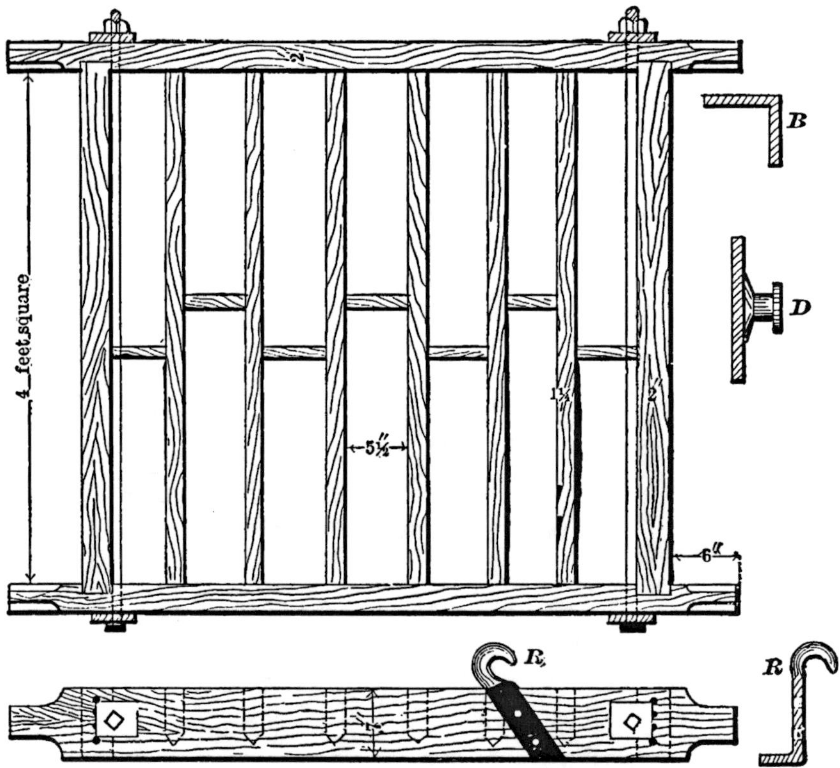

at any time liable to get a job to make, for which he will be obliged to take parts of several flasks—perhaps two or three copes and as many bottoms—and bolt or nail them together. Quite likely, to get what he wants, he gets in parts of different flasks, and another moulder discovers that some of these pieces are just what he wants, and eventually, instead of being put where they belong, they are dumped down indiscriminately at the handiest place. Perhaps, to complicate matters, the foreman will order some of the parts of a half dozen flasks put carefully aside for future prospective use, on another piece that *may* be ordered. In any event, the parts of different flasks get promiscuously mixed, and the result is that some moulder looking for a cope to cover a pattern bedded in, or wanting a bottom to raise some part higher, will see these parts so nicely piled up, and the part that he wants at the bottom of the pile. Sooner than go for the man who has charge of the flasks to help him, he will throw down the whole pile, smashing pins and handles. This is the way the thing works in almost every regular jobbing foundry.

A foundry for special work does not require much over half the room that a jobbing foundry does, for storing flasks. In a foundry designed and equipped for special work, there is generally a large number of flasks for the same pattern; and when any pattern is brought into the foundry to be made, the foreman can send for the required flask, without spending two or three hours' time in looking for stray copes and bottoms. Their flasks can be piled up high, because they do not have to be disturbed to get a part of a flask from the bottom of the pile. In such shops men will work months—sometimes years—with the same set of flasks, whereas, in some jobbing shops, two weeks would be the limit.

In a jobbing shop, as a rule, the moulders, when going to work in the morning, have no idea what job they will start on, or the number of new jobs they may be called on to make before getting home again.

An assortment of miscellaneous jobbing flasks should have plenty of ground room, and the piles should be made open and not very high, so that a man looking for a flask will be able to see every part without having to throw down a pile. There should be some one in charge of the flasks, and he should report any man known to ill-use them, and then, if the foreman does his duty, there will be no need of moulders losing their night's sleep worrying over a drop-out caused by loose bars, crooked pins, or shaky flasks. Of course this does not mean that a jobbing shop cannot be run without having an acre or two to pile up flasks on. There are plenty

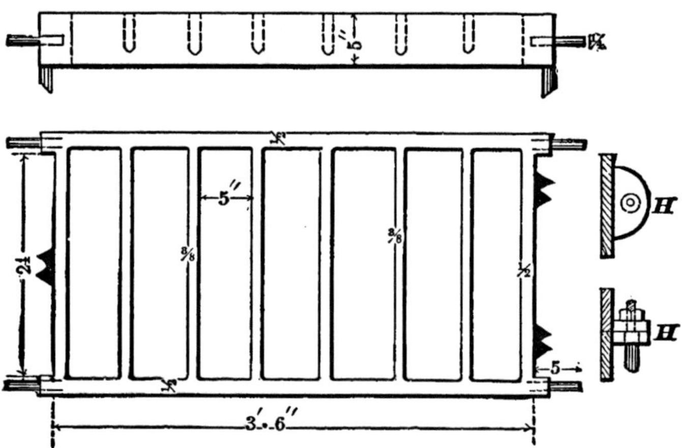

of shops doing a large business that do not have much ground room, but have to hoist their flasks on the top of flat roofs; their very large ones are never taken out of the shop.

In this country, wooden flasks are used more than in foreign countries. When foreign moulders come to work in America, they wonder at so many wooden flasks being used. They are sometimes afraid to start to work with them, and if they have a run-out, or drop-out, the old wooden flasks get the blame.

Wooden and iron flasks each have their especial advantages : the iron one is the most durable, is stiffer, and can be relied on in matters of dropping, run-outs, or strained castings. Take a large plate full of holes, cast one with a wooden, and one with an iron flask, and then look at the difference in the two castings. The one made in the wooden flask, with all the time spent wedging down the bars, and blocking up to get the weights, or screw-down binders on, may be from $\frac{1}{4}''$ to $\frac{1}{2}''$ thicker in the middle than it should be, and as for the holes, they are not to be seen on the cope side at all.

Wooden flasks have also their advantages. They are handy to lift and carry, and better adapted for a dull foundry hatchet, to carve out the bars, so as to admit of the cope being used for various castings. After being pitched around eight or nine times, they will also save the cupola man lots of time and labor in hunting up kindling-wood to start his fire with.

There are more iron flasks used in this country at present than were formerly used. Some shops that have a standard of work can rig up iron flasks with loose bars and side pieces, that can be bolted together to answer the purpose for making a variety of castings.

The sketches represent the principle of construction, etc., of plain iron flasks, such as every jobbing shop should or could use. The smallest ones, without bars, are very light and handy for small jobbing castings. The second size is a cope about as heavy as two men can lift off. The lug and pin, *H H*, should be attached to special flasks where parts of the casting are to be made in the copes. This pin is expensive, as it involves considerable machine-shop labor to make, as in turning the pins and boring the holes, but it will pay for itself in a short time, when used for making pulleys, etc.

For dry sand work that requires to be closed very true, the flasks should have three or four pins. The pins may sometimes need to be one foot in length, so as to close true over cylinder port cores and the like. The pin shown, cast on the small flasks, costs very little to fit up, and is very good for a plain class of work.

The large iron flask shown possesses several advantageous

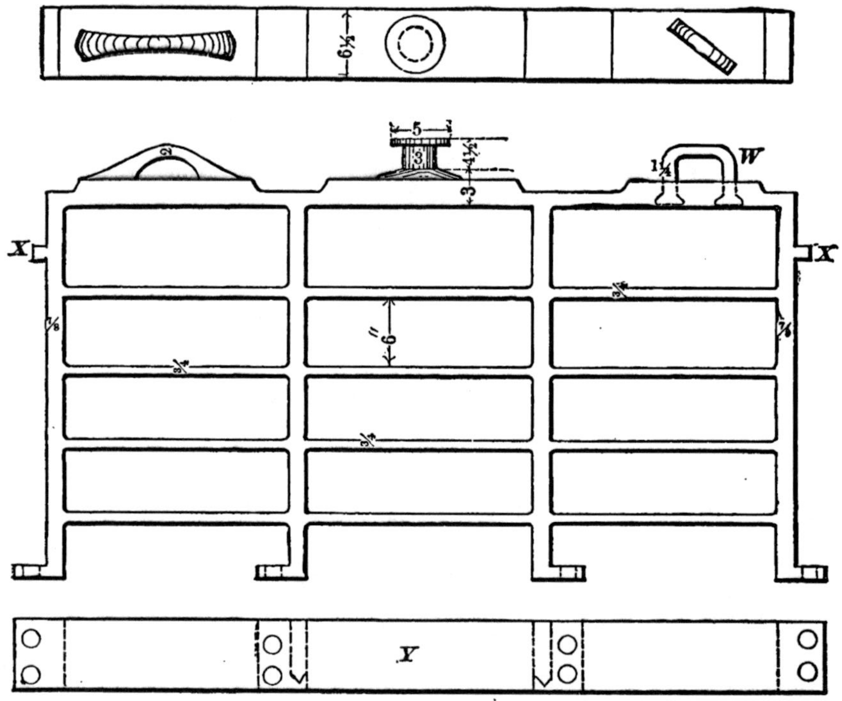

features. The round trunnion is a common but very handy thing for turning over copes, so as to get to finish them easily. This may be cast on, the flask having handles, also, if required. The handle, *W*, is of wrought iron, cast into the flask, which makes a neat lifting-handle. It is cast in on a slant, so as to be in a line with the chains when lifting. The handle on the opposite side is cast-iron, and should be large for strength, but should be made convenient for hooking to. Guides, as shown at *X,X*, should be cast on for driving down stakes alongside of. The loose plate or bar *Y* can be bolted to the flanges, should the flask require to be made longer, or to have a piece or pocket bolted on for any purpose. To accomplish the same purpose, the whole flask is sometimes cast together, and one end cut out about $\frac{3}{4}''$, so there will be no bearing on the joint, and when the flask is wanted longer, a section is bolted against its flat surface. The objection to this is, that the end cut out is never solid on the joint when the flask is used without the extension, and when there is a piece bolted on it, the joint forms a flat clumsy surface for the sand to hang to.

The flasks shown are intended for a plain class of green sand work, such as almost every jobbing foundry has to do. Deep flasks, in some instances, are better for being made in sections, and bolted together. Iron flasks for dry sand work are better if made $\frac{1}{8}''$ thicker than shown in the cuts, as they have to stand rougher handling.

The cut of a wooden flask shows a good reliable way of making a stiff flask for a medium class of work. The angle piece *B* is cast-iron, and is a good thing to put in the corners of large flasks for bolting the sides together.

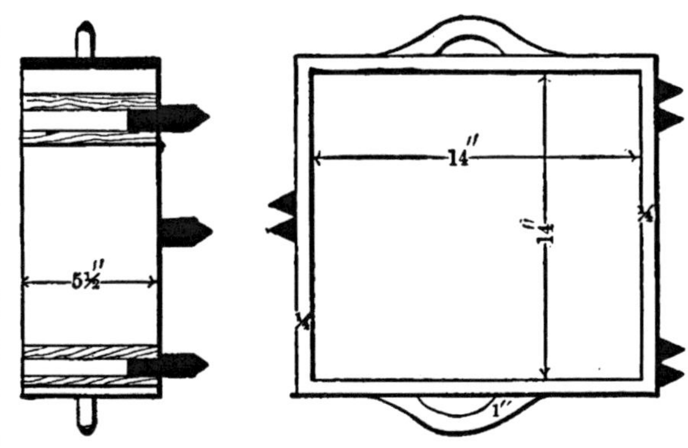

This angle would be better if a small bracket was cast on the inside corners, so as to make it stronger. *D* shows a trunnion that can be bolted to a flask, to roll it over, or the two ends could be entirely of iron, with the trunnions cast on, and wooden sides bolted to them. When bolts will not hold a large flask stiff enough, cast-iron bars are sometimes used instead of wooden ones, having flanges cast on to bolt the sides and bars together. The handle *R* is wrought iron, to come under the bottom of the cope, and has two bolt-holes in it. This makes a reliable lifting hook for very heavy copes.

Should a moulder wish to know the weight of the sand he has rammed up in a flask, in order to tell if the crane or chain is strong enough to lift it, he can remember that one cubic foot of sand when rammed, and of the right temper, weighs about one hundred pounds.

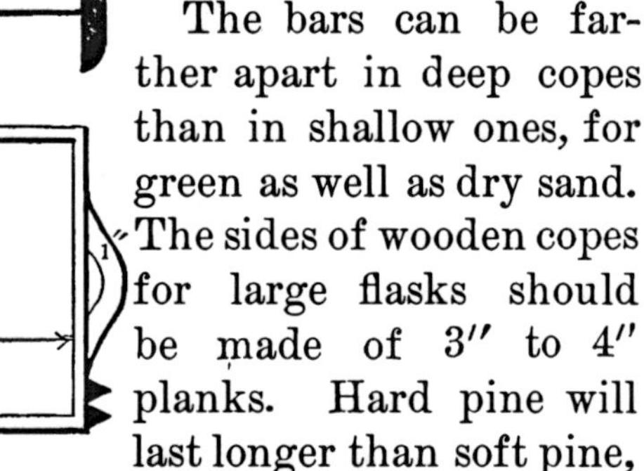

The bars can be farther apart in deep copes than in shallow ones, for green as well as dry sand. The sides of wooden copes for large flasks should be made of 3″ to 4″ planks. Hard pine will last longer than soft pine, especially if pounded much with a sledge-hammer, which should never be done if there are wooden mallets to be had. In making iron flasks the best iron should be used, as a good flask is an essential feature in turning out good castings. They should receive care and proper handling, otherwise, in a short time, a new flask will be only fit for kindling-wood or scrap.

SKIMMING AND FLOW-OFF GATES.

As melted iron has more or less dirt or impurities, which keep rising upon the surface of the metal, more especially while it is exposed to the air, it is of the utmost importance to have the runners and gates made so as to collect the dirt as much as possible before the metal enters the mould, to insure a clean, solid casting. The gate shown is an improvement on the common skimming gate, as there is one more riser or dirt-catcher in it, into which the iron goes circling round, whirling the dirt up to the top. It is astonishing how little thought some moulders have about the principle of skimming gates. Go into almost any foundry, and you will see men making or working on good work, setting the largest runner for the pourer, and the smallest for the dirt-catcher; or they will cut the gate that goes into the mould larger than any other portion of the runners or gates. I have also seen skimming gates cut when the man cutting them did not know which one of the upright runners was the one to pour into. This showed that, of course, he had given no thought to the subject. There were, in the instance referred to, two upright runners, with a channel cut between them, and he thought he was cutting a skimming gate. In the accompanying cut is shown a crank for an engine, bedded in the floor. To save work, the face is cast up, and it requires the greatest of care as regards clean iron going into the mould. The gate that leads into the mould is cut the smallest of any, so that the rest of the gates and runners may be kept full of iron. The dirt flows up to the top of the risers

A and *B*. In this way clean iron goes into the mould. Risers *A* and *B* have no connection cut in the cope part, it being cut in the bottom, from *B* to *A*, and on a circle, so that the iron will whirl around in the riser *A*. The runner and pouring gate *D* are connected with *B* in the cope part, but can be connected in the bottom part, like *A* and *B*, should it not be practicable to connect in the cope. The pouring runner *D* is large enough to keep the lower gate full of iron, and the two dirt-holders or risers are larger by one-third than the pouring runner.

The pouring basin *M*, if for a very heavy casting, could be made longer, and a skimming core added, as shown in a previous article on "Making a Heavy Green Sand Casting." All the runners should be rammed even, so that there are no soft spots in them, and all corners or edges of the gates and runners made rounding, so that the running iron can have no chance to wash sand into the casting.

The numbers 3 and 4 show a good plan of risers to take the strain off the mould when pouring. The riser 3 is connected with the pouring gate, and a clay ball stops the iron from flowing away until the mould is full. It then flows down the outlet, under the joint of the flask. The connections between 3 and 4 can be cut down as low as 3 inches, which leaves very little strain on the mould. This is also used independently; but cutting the riser from the mould, and having three or four of them, causes a sudden pressure on the cope to be greatly released. As a good skimming gate is essential in making a clean casting, so are good risers necessary to keep a casting from being strained. In this respect they are of equal importance, and too much attention should not be given to one to the exclusion of the other.*

* The question of skimming gates and procuring clean castings is an important subject. To thoroughly understand the subject, the reader must refer to the chapters found on pp. 16, 39, 48, 114, 120, 129, Vol. II.

D B A

3 4

LONGITUDINAL SECTION

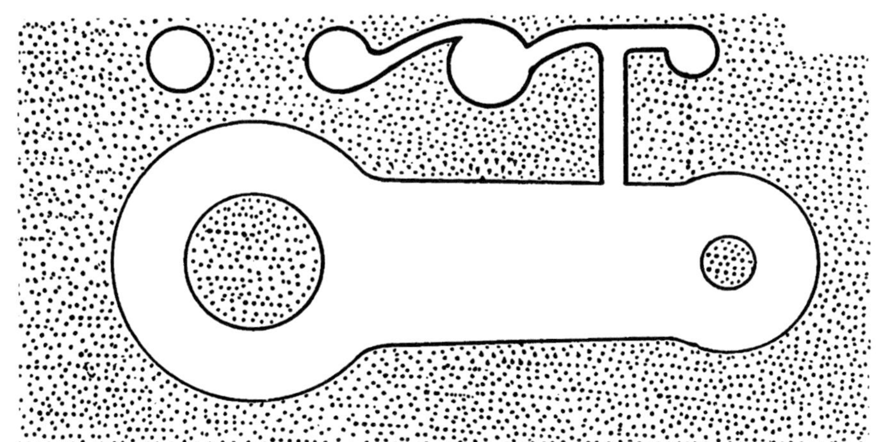

PLAN & SECTION

MAKING A GREEN SAND BASIN—RUNNERS AND GATES.

In making castings, the basin, runners, and gates are often responsible for their being bad. There is little in the whole art of moulding that requires more care than the making of these parts of a mould. A moulder may slight the rest of his mould and have his casting come out all right, but any carelessness or ignorance in making the basins, runners, or gates, will almost always cause trouble

In pouring a mould, the iron first drops from the ladle into the basin ; from the basin it runs with more or less of a rush into the upright runners, from the runners into the gates, and from the gates into the mould. With the exception of that portion of the mould which the iron enters or drops into, there is very little agitation of the metal as it gradually rises.

In the cut shown, *H* is the cavity into which the iron first drops as it is poured out of the ladle; *Y* is the runner through which the iron flows to the gate *K*, from which it runs into the mould.

For pouring five-ton ladles, the width of a basin should not be less than 18″, and the depth should be 9″. The bottom of the basin, where the iron first drops, should not be less than 2″ deeper than at *S*. From *S* down to the runner, *Y*, there should be an easy incline ; the longer the basin the more incline there should be.* This assists the iron in flowing, making sure of keeping the runners full, and also prevents any iron remaining in the basin when the mould

* This subject is further discussed on p. 130, Vol. II.

is full, except that which is in the cavity that is formed for the iron to drop from the ladle into. This cavity is provided for the purpose of preventing the cutting of the bottom of the basin, which would be the case was this part made even with the rest. This cavity, which is soon filled, allows the iron to fall into iron instead of on sand. When this cavity is filled, the iron runs easily from it to the mould, lessening the danger of cutting, and allowing the iron to be rushed in, so as to keep the runners full.

For the pouring of larger ladles than five tons, the width of basins should be from 18″ up to 30″, and the depth from 9″ to 15″.

When making green sand basins, the sand should be well mixed and riddled before it is shoveled into the basin box. The careless use of unmixed sand for making basins often causes bad castings.

There is one way of making basins that many moulders follow, but which a careful moulder will never employ; that is, they will shovel in some sand, and form the shape of the basin by packing up the sides with handfuls of sand. This makes a loose basin, and one that is liable to cause trouble.

To make a reliable basin, the box should first be evenly rammed full of sand, after which the shape of the basin can be dug out with a shovel or trowel, thereby giving a firm, solid basin, as far as the ramming of the sand is concerned. The trouble with basins usually commences at the bottom, and is caused by the falling iron cutting the sand or letting the iron get to the wooden bars. A good moulder will never have less than 3″ of sand between the bottom of his basin and the wooden bars.

There are two or three ways that the bottom of green sand basins may be secured so as to prevent cutting when used to pour heavy castings. The first is to stick nails all over the bottom, having the heads even with the surface of the sand.

The second and third ways are to set in a flat core, or two fire or common bricks to form the bottom. In either of the last ways there is very little danger of the falling iron causing any trouble.

For the pouring of heavy castings it is best, when possible, to build basins outside the copes instead of on the top of them, for the reason that high heads can thus be avoided, thereby not having so much strain upon the mould. The cut of basin shown is one thus made. In this basin will be noticed a coke or cinder bed placed underneath the basin, which is a very good plan for large surface basins, as it will carry off the gases and steam, and thereby prevent any boiling or scalding of the bottom of the basin. Sometimes dry sand basins are made, and hoisted from the oven carriage and placed where wanted for the pouring of very heavy castings. Some moulders prefer to use them instead of green sand basins.

Another part of a green sand basin that often gives way while the mould is being poured, is the front, *X*. Sometimes a moulder will cut out the front as if he were trying to make the end at exact right angles with the sides of the basin, as shown at *P*. This may do well enough for small basins, but for large ones it should never be done. The safest way is to form this end as shown at *W*, having the end very nearly a circle. Sometimes it is best, in the instance of very large basins, to have this end of the box full of nails, driven so as to stick out two or three inches, to have a good hold of the sand; that is, when wooden basin boxes are used, but for iron boxes (which, when possible, should be given the preference), the front should be secured by being roded.

How often have moulders seen castings lost by having the wooden basin box spread open, which would have been prevented had there been some narrow strips of wood nailed across the bottom, as shown at *B;* or cast or wrought

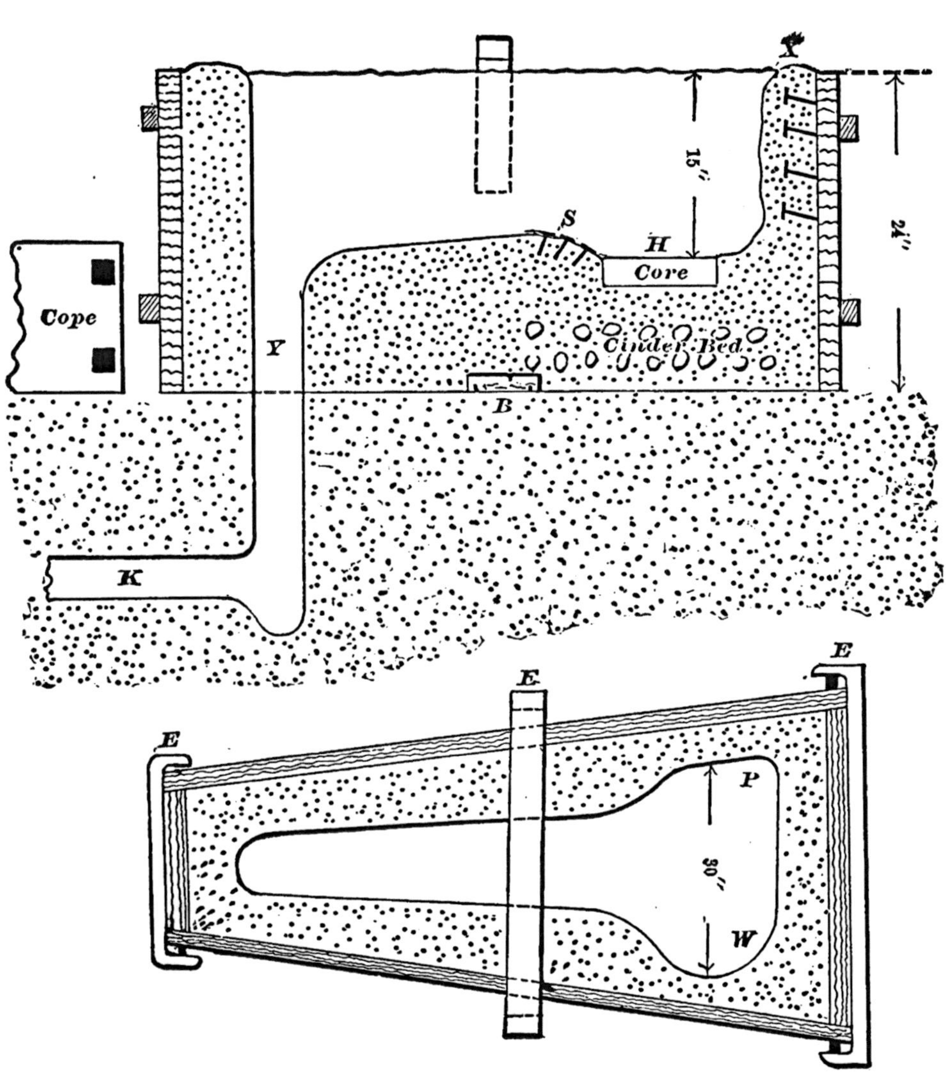
X
15"
24"
S
H
Core
Cope
Y
Cinder Bed
B
K
E
E
E
E
P
30"
W

iron clamps used to hold the sides together, as shown at *E E E*.

Again, the iron will burst out from underneath the box, from the lack of weights or wedges to hold it down

There are very few moulders that can be trusted with the making of a basin to pour a large casting with. They *think* they know, and it is not until they have lost a number of castings that they are convinced of their carelessness or ignorance. They are not always convinced, but will lay the blame to the basin box, the sand, the helper, or will confidentially tell some of their friends they were made to pour the iron too hot, or too fast, and that no basin would stand such treatment.

The rammer and swab pot are very necessary tools in a foundry, but in the hands of a thoughtless or ignorant moulder they are about as dangerous as a loaded revolver in the hands of a child. There are no tools used in a foundry that are more responsible for bad casting than the rammer and swab.

"Bring me that rammer," yells the moulder to his helper, and when he gets it he uses it lustily.

"Bring me that swab pot." He gets it, and on goes the water, plenty of it, too. What's the use of being afraid of water?

Around comes the ladle, and out of the ladle into the basin goes the iron. From the basin up to the roof it flies, the men let go the ladle, and run to the corners to see if they are burned; then they will sit down and think of the moulder and his swab pot and rammer.

When a good moulder, or one that thinks he is, loses a casting on account of his basin, he should never blame any one but himself; for he should know from experience what is required.

A moulder can tell by the looks what sizes of runner

sticks or gates he wants. If a moulder orders a runner or gate stick made without having one to get the size from, he will hardly ever be satisfied with it.*

In thinking of the number and size of runner or gate sticks for a mould, there are many points that should be considered. The first is the weight of the casting, and whether it should be poured fast or slow. The second, the form of it, and whether its proportions are heavy or light. The third, what temperature of metal will the mould require to be poured with ; will it be run from the bottom or from the top of the mould; and also the height of the basin above the mould.

As a general thing, the faster a mould can be poured with safety to all of its parts, the better it is. It is not always the weight of a casting that decides whether the runner and gate sticks should be large or small, to pour the mould fast or slow. The common, old style grate-bars, that have thin openings or cores in them, are a good example to show why some moulds require to be poured slowly.

Many very good moulders have worked on this class of castings, and have been astonished at their lack of success in making good ones. A grate-bar, or any mould that has similar thin green sand cores in it, should be poured with hot iron, and slow. Pouring them slowly gives the gases and steam in the sand a chance to escape through the vents, and the iron, being hot, will easily run level, and not pile up higher nearer to the gates than in distant portions of the mould, as dull iron will generally do, thereby causing thin bodies of sand to be displaced. Hot iron will also admit of the pouring of such moulds slow, without danger of the castings being cold shut.

Moulds having thin green sand or dry sand cores in them, should often be poured hot and slow for the reasons above stated. Very heavy castings that have no dangerous dry or

* A valuable table upon the subject of gates is found on p. 244, Vol. II.

green cores in them, should have large runners and gates, so as to admit of the iron being poured dull and fast, and the same with any moulds that the copes would easily draw down.

There are also large and small moulds that require hot iron poured in them fast, in order to have all the parts run and not be cold shut.

Again, there are some moulds that require the liquid iron to be forced into them as fast as possible, for which high heads or basins should be made. Castings, such as cylinder or pipe-shaped moulds, that are cast vertically, should have larger runners and gates where they are poured altogether from the bottom, than where they are poured from the top, by dropping the metal down, as when the iron is all poured from the bottom it gets duller as it rises up in the mould. Many times, such moulds are poured from the bottom and top also, so as to avoid having any trouble from dull or dirty iron in the upper portions of the casting. A cylinder, etc., will bore out cleaner if poured from the top, than if poured from the bottom.* Iron dropping from the top keeps cutting up any forming lumps of dirt, causing it to float and keep on top of the rising iron ; and also when iron is run from the top, there is as hot iron in the upper portion as in the lower portion of the mould. But when poured altogether from the bottom, the iron becomes dirty and duller as it rises up, and the dirt will collect in lumps and roll under flanges, cores, etc., and also lodge against the sides or surfaces of the mould, and it is not until the casting is bored or plained, that the dirt is seen. Iron should always be poured into a mould as far as possible from the parts that require to be finished up the cleanest, since the dirtiest portion of a casting is where it is poured or gated. A smoother skin can be made on a cylinder by pouring it from the bottom, but it will be at the expense of its being dirty when bored.

---

* For a thorough discussion upon procuring clean cylinders. see pp. 38, 48, Vol. II.

WEIGHTING DOWN COPES—DAMP FOUNDRY FLOORS.

MELTED iron supports and floats a body the specific gravity of which is not greater than its own, the same as water or any other fluid. Solid cold iron floats on the top of melted iron, similar to ice floating on water. Water, when frozen, expands, but the expansion does not make the body any heavier or lighter. A definite quantity of water weighs the same whether liquid or frozen. Water, in changing to ice, expands about one-ninth of its bulk, which makes ice specifically lighter than water, and therefore it swims or floats on it, about eight times as much being below as above the surface.

An iron ship sinks until it displaces water equal to its weight and then floats; but if the same quantity of iron were in a solid mass, it would instantly sink to the bottom.

In each of the above cases the cause is quite plain; but in the instance of iron floating on iron, the matter is not so apparent. With ice there is an observable expansion, which in solid iron is not seen.

With reference to the floating of water and iron, the iron, when melted, must be specifically heavier than it is when cold; or iron, when cold, must be more bulky than when hot, in order to be the same as ice. But to say that iron expands so as to occupy more space when it is cold than when it is melted or hot, would be to ignore observable results that occur in almost every casting made. When a pattern is constructed to make a casting from, it is made from $\frac{1}{20}$ to

$\frac{1}{8}$ of an inch per foot larger than the casting is wanted. In an open sand mould for making a bar, ring, or plate casting, after being poured, the liquid iron begins to cool and contract, and the contraction is steady and visible from the beginning to the end.

Cast iron expands at the moment of solidification. This is one reason why heavy casting, after the mould is filled, requires so much feeding, taking from 100 to 400 pounds of melted iron to supply the shrinkage of the cooling iron. A study of these questions is worthy of attention and experiments.

There is one thing that moulders know to be a fact, and that is, melted iron, when poured into a mould, will raise a cope so as to run out, and perhaps make a bad casting, if the cope is not bolted, clamped, or weighted down sufficiently to resist the head pressure, and the momentum with which the rising iron comes up against the cope or covering.

The weight a cope requires on it, to hold it down, depends on three distinct things; the first of which is the height of head or pouring basin above the casting; the second, the velocity with which the iron comes up against the surface of the cope; and the third, the number of square inches in the lifting surface of the cope or core part of a mould.

In every mould these three conditions, in a greater or less degree, are present, and must be provided against. The momentum of the iron against the cope is the reason why a flask sometimes requires so much weight on the cope to hold it down. If it were not for this sudden pressure, as it were, one-half of the weight used would in some cases be sufficient.

For a moulder to say, as some do, that he has a standard rule whereby he can figure up the pressure on a cope, and hence *obtain the exact weight* required to hold it down, may mislead some. For one to say he can figure *exactly or just sufficiently to hold down copes*, is to claim that "weighting down" calls for the *consideration of no conditions*, and that the pressure copes receive is purely the metal's "statical head." Were this true, one could easily figure exactly the weight *just sufficient* to hold down copes. In order to *safely* "*weight down*," there must be more weight as a general thing used than the "statical head" would figure. The reason for this is mainly due to the momentum *pressure* copes receive. In some cases it will be very light, while in others it may be so great as to call for near as much again weight as the "statical head" would figure.

As there are so many features to be considered and understood before rules could be safely given for "weighting down," the author has deferred the matter to Vol. II., where, commencing with p. 187, a very large practical chapter upon "weighting down copes and cores" is given in a manner the best moulders will concede as being original and progressive.

The cuts *B* and *W* show in plan and side view a cast-iron weight of about one ton, made expressly for weighting down flasks. In a shop provided with a number of these, a cope can be reliably weighted very readily. The *V* grooves are cast in the weight, so that it can be broken and remelted whenever desired. Some shops, by their own bad castings, have supplied themselves with weights, while some buy heavy scrap, and again others will fill wrought-iron rings with pig-iron to make crane weights.

The cut, showing a cope bolted down, represents a plan that some foundries have adopted to save labor in hoisting on and off weights, and to insure safety. It is almost impossible for a cope to rise so as to have run-outs when firmly bolted down in the manner shown.

One, or as many bolting-down floors as can be used in a

shop, will save time in making castings that have great lifting surface on the copes.

To make a bolting-down floor, as shown, a hole is dug as deep as required, and bottom cast-iron binders placed solid and level.

On some floors that are wanted for long or large moulds, there can be as many binders as required, set a handy distance apart, which, for a common run of work, is about four feet. The binders should be about twelve feet long. On the top of these binders are bedded some heavy planks, and the sand shoveled in on them and rammed solid ; then the bolting-down floor is ready for use.*

The slanting coke or cinder bed, shown below the upper bed, is sometimes good for moulding castings in a wet or damp floor, which, however, is a very bad thing to be bothered with, as there is nothing that a moulder dreads worse than having to make a deep, heavy casting in a damp floor.

I have made castings by this plan where the lower part of the floor was nothing but mud—sometimes so wet that iron plates were laid down before the slanting coke-bed could be made.

By having two beds, as shown, the vents and heat from the melted iron, when poured into a mould, is taken off by the upper bed. If the heat should get to the lower bed, so as to cause steam, there might be danger of scabbing or blowing were this lower bed the only one used ; but, as shown, there is no danger. A deep wooden box, *X*, is set in, so that if any water collects below it can be bailed out.

Some shops can be sewered so as to drain off the water. But if this is not possible, it is a good plan to have a deep pit or well dug in a shop to collect the drainage, and to collect the water from wet floors.

Floors could have plates full of small holes under them, resting on the top of flat timbers, placed three feet apart.

---

* By referring to pp. 204, 229, Vol. II., further information upon bolting down floors will be found.

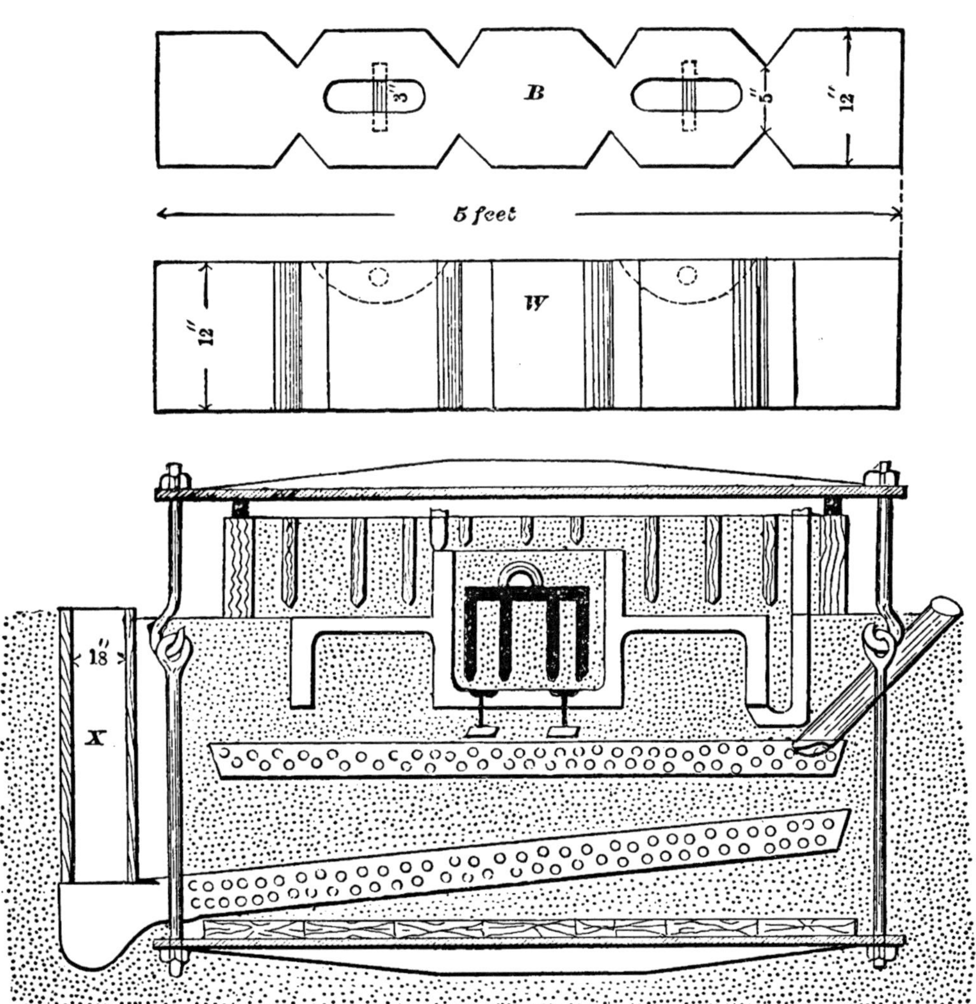

WEIGHTS FOR FLASKS—PLAN OF MOULDING ON WET FLOOR.

The timbers should rest on large flat plates on a solid bottom. The timbers between the upper and lower plates will leave an open channel under the mould, and with a good bed of coke or cinders laid on the plates, any water collected will pass through the holes in the plates to the large open channels below, and run into a well, from which it can be bailed or pumped out by hand or by a steam-pump.*

Some shops have large tanks of boiler iron sunk into the floor, and by moulding inside of them they avoid the dampness. But there is much objection to their use, on account of the shop room spoiled by the edge of the tank coming in the way of general jobbing work.

---

* As a general thing it is only where there is sandy soil that foundries will experience trouble from damp or wet floors. Where foundries are located upon *deep strong clay* soil, there is generally no trouble from damp floors or water collecting in pits, etc.

ONE HUNDRED ITEMS THAT APPRENTICES SHOULD KNOW AND REMEMBER.

1. MOULDING sand contains gases.

2. The gases in moulding sand pass out when iron is poured into a mould.

3. These gases may pass off in a right way or a wrong way.

4. When iron bubbles or boils, the gases are passing off the wrong way.

5. When gases pass up through the iron, it is because the sand is rammed too hard or is not properly vented.

6. Hard ramming closes up the porosity of the sand, but the vent wire opens it.

7. Too much water used in mixing sand creates too much steam when the mould is poured.

8. Steam accumulating under liquid iron, will raise and blow it up.

9. When iron bubbles or boils in a mould, it will make a scabbed or bad casting.

10. Iron naturally seeks a soft bed. If poured on a hard bed it will bubble and fly.

11. The face of a mould should always be the softest.

12. Whenever hard ramming is required, it should be done on the outside portion of a mould.

13. A rammer should never be allowed to strike a pattern when ramming up a mould.

14. When ramming up the sides of a pattern, the pin should not go nearer than one inch to the face; the butt should be kept one and a half inches from the face.

15. Too much depth of sand in a ramming will be apt to make swells upon the sides of a casting.

16. In ramming courses of sand, the rammer should be made to feel the under or last course rammed.

17. Too much venting will seldom do any harm, but hard ramming will.

18. Hard ramming requires good venting.

19. To vent hard ramming requires muscle, and do not be afraid to employ it.

20. Learn to ram even and lightly.

21. Some moulders will ram harder than others.

22. He that practices hard ramming will always have the most trouble.

23. Some parts of a mould require and will stand harder ramming than others.

24. The higher the head or pressure upon the lowest portion of a mould, the harder should be the ramming there.

25. Plain vertical sides of a mould will stand harder ramming than the flat bed-surface of a mould.

26. Cores, or a projection that is surrounded by iron, should be rammed even and lightly, and also well vented.

27. Any bottom section of a mould that is covered rapidly with iron, so as to have a pressure upon it, will stand harder ramming than where it is to be covered over slowly.

28. The lowest point of a mould's sides is the one which should be rammed the hardest.

29. The highest point of a mould's sides is the one which requires to be rammed the lightest.

30. The flat surfaces of copes will stand harder ramming than the bottom surfaces of the mould.

31. In plain copes the gases and steam are on top of the iron.

32. In beds or bottoms of moulds, the gases and steam are underneath the iron.

33. Gases or steam do not act well underneath hot iron.

34. The bottom portions of rammed moulds should always be provided with ways and means to let the gases and steam escape easily.

35. If there is not a good opportunity for gas and steam to escape downwards, it is seldom that there is pressure or body of iron strong enough to keep them from passing up through the iron, and thereby causing a casting to blow or be scabbed.

36. It does not require so much pressure or force to drive steam or gas upwards through iron, as is required that the iron may hold or force steam or gases downwards.

37. The harder the ramming, the more force or pressure will iron require in order to drive the gases or steam downwards.

38. There is less gas in old sand than in new.

39. Facing-sand, or sand having blacking, flour, or sea-coal mixed in with them, contain an increased amount of gases.

40. The more gas there is in sand, the more venting should be done.

41. Avoid using the swab as much as possible when finishing a mould.

42. Finishing a mould by often using a swab, makes rough, scabby casting.

43. Never patch a mould with a trowel, when you can patch it with your hand or fingers.

44. The less sleeking done in order to properly finish a mould, the better will the casting be.

45. Never sleek twice where once will do.

46. Patching a mould with your fingers will never cause a scabby casting, but too much sleeking patching will.

47. When finishing, the lighter you can bear upon your tools the better.

48. Heavy sleeking closes up the pores, and makes the surface of the mould hard.

49. A hard surface sleeked mould is apt to cause cold shut, and thin scabs on a casting.

50. The dryer sand can be worked and practically used, the better.

51. The dryer the sand, the more ramming it will stand.

52. The more ramming a mould will stand, the more strain can be put upon it. For heavy castings these last three numbers should be specially remembered.

53. A mould that needs to be poured fast, generally should be well vented.

54. A fast-poured mould should be well made.

55. The best way to know if a mould is a good one, is to fill it with iron, and then see if the casting is perfect when it is cleaned.

56. It is not necessary that patterns should be jabbed full of holes in order to vent a mould.

57. One inch thickness of facing-sand, over a mould, will peel a casting as well as a foot thick.

58. The use of facing-sand has caused more bad casting than common sand.

59. Using the facing-sand too strong causes cold shut, or streaked castings.

60. A cold shut casting is harder to deal with than a scabbed one. Scabs can be chipped off, but to hide cold shuts has puzzled many a moulder.

61. Do not use facing-sand 1 to 8, when 1 to 10 would peel the casting.

62. Facing-sand will stand the wash, or running of iron, better than common sand, where gates or runners are cut.

63. The first course of ramming in a cope should be evenly and firmly done.

64. The second and third courses do not require as much careful ramming as the first does.

65. The butt rammer is used to make the sand solid between the bars, so as to keep it from dropping out.

66. In using the butt rammer, be careful not to pound the cross bars, as you will be apt to loosen the sand on the face of the bars.

67. Always know that the bars of your flask are solid and nailed in, so as not to get loose.

68. When your cope is on, try if it will twist in the pinholes or not, as a casting that has been made in a mould that has not been carefully closed, looks badly.

69. Whenever you cut out the bars of a cope, do not for get that bars in a cope are put there to hold or lift up the sand.

70. If you can avoid having gaggers sticking up above the top of a cope, do so, as by this many a casting has been lost.

71. Avoid using thick, heavy gaggers as much as possible.

72. Always make sure of having two-thirds of the length of a gagger between the bars, when you have a body of sand to lift.

73. The sand between the bars holds the gaggers, and the sand below the bars should be lifted by the gaggers.

74. If there is not sufficient depth of sand between the bars, exceeding in inches that below it, wooden soldiers should be tacked on the bars to assist the gaggers.

75. Never lose sight of the fact that iron gaggers are heavier than the same proportion of sand.

76. A hanging body of sand would stand a better chance of being lifted by having no gagger, than by having a lot of short gaggers coming up but a short distance between the bars.

77. Keep the top of your pouring runners free from a lot of loose sand around the hole, by packing it firmly with your hand, and then swabbing it over lightly with water.

78. Never pour a ladle of iron unless it will be skimmed, and stand so that you need not move after commencing to pour.

79. If the castings are light, they should not be left in the sand over night, as they are apt to get rusty.

80. Castings that will keep red-hot for two or three hours after being poured, are best kept covered over with sand until they become cold, since leaving them exposed to the effects of the atmosphere, destroys the good color of a casting, and its strength.

Loam and Dry Sand Moulds.

81. Keep on good terms with the foreman, if you wish a chance to learn these branches of the trade.

82. Dry sand moulds, as a general thing, should be rammed harder than green sand.

83. Dry sand does not require as much venting as green sand, and there are many moulds that can be cast without having a vent in them.

84. There is less gas in dry sand than green sand, and if a mould is thoroughly dried, there is no steam to contend with.

85. When ramming dry sand moulds, be just as careful to avoid hitting the pattern with the rammer, as with green sand moulds.

86. All joints made on dry sand or loam moulds must be pressed or sleeked down, so as to leave a fin upon the casting, as this class of moulds would crush, if the joints were left as the mould was parted.

87. There is an old maxim, that it is better to have a fin than a crush, the truth of which many old experienced moulders have found to their sorrow.

88. When finishing dry sand moulds, sand should never be patched or sleeked on smooth or sleeked surfaces.

89. Dry sand mixtures depend upon what kind of sand or loam a shop uses; almost every shop has a different way of mixing dry sand and loam.

90. A close mixture of loam or dry sand is very liable to scab, while if it is too open, the mould will not stand the dropping or washing effect of the iron, when poured.

91. To know what proportions of sharp and loam sands to mix together, a man must have experience, but some few places have a natural loam or dry sand that requires no mixed proportions.

92. Dry sand or loam moulds, if not thoroughly dried, generally cause a casting to become scabbed.

93. A casting poured hot will finish up cleaner than one poured dull, providing both are free of scabs.

94. It will take more strain to break cast-iron when there is heat in it, than when it is cold, and the same is true of wrought iron. In winter, or very cold weather, chains should not be used to hoist as much as in the summer time; sometimes it is best to heat a cold chain before hoisting a heavy weight.

95. It is not safe for two-ply crane chains to hoist any more than—

20	tons	with a	1″	chain.
14	"	"	$\frac{7}{8}$″	"
10	"	"	$\frac{3}{4}$″	"
6	"	"	$\frac{5}{8}$	"
3	"	"	$\frac{1}{2}$	"

If chains are not made of the best iron, we should not hoist more than two-thirds of the weights given.

96. When learning your trade, don't let your conceit run off with your common sense, as such conduct makes your superiors dislike you, and also hinders your advancement.

97. Whenever you want to know anything, if you have friends, they will tell you, and without such help you can do but little.

98. Apprentices cannot afford to lose the good-will or friendship of any one in their shops.

99. A good apprentice will make a good journeyman.

100. Never allow yourself to think that you have learned the entire moulder's trade, for one's knowledge here can constantly be increased; no man has yet mastered the moulder's trade.

BUILDING AND FIRING LARGE OVENS.

THERE is nothing in a foundry that is ordinarily so illy constructed, and with which so much fault is found, as the ovens for drying moulds and large cores. Ask any moulder who has traveled considerably, how many ovens he knows of that give good satisfaction, and it will tax his memory to tell of more than two or three, and then if they are not located so as to be in the way, or take up the best portion of the shop, it will be a wonder.

When building a foundry, the locating of the oven should be attended to by a thoroughly practical man, and is a matter that should receive much thought and attention, as there are few shops in which they can be built on the same general plan. Ovens should be built where they will be out of the way of doors, gangways, and green sand floors, and, if possible, should be in that section of the shop where the loam and dry sand work can be done to the best advantage. In some shops ovens are placed so that the track has to be curved in order to run the carriage under the sweep of the crane. This is a very bad plan, as the car, when heavily loaded, moves hard on a curve.

When the ground room will allow it, and the shop room is small, it is best to have the ovens built outside the shop, having the entrance even with the inside wall.

In building ovens it is also important to know and provide for the class of work intended to be made or dried, as some work will not stand to be dried fast. Such work as large cores, cylinders, gears, or any fine dry-sand or loam moulds, should be dried with a fair, even fire, especially if the moulds are to be blacked dry. Such moulds as rolls, spindles, propeller-wheels, or other coarse work, will usually stand a hot fire. There is nothing so bad to handle as burnt moulds or cores, for which the poor night-watchman seldom escapes blame.

Mixtures of dry or core sand, having plenty of loam and sharp sand in them, will stand a hotter fire without being burnt than sand having flour, meal, or much moulding sand mixed with it.

For large cores that must be dried quickly, and without cracking or being burnt, the less flour and the more clay-wash used for mixing the sand, the better.

For fuel in firing up ovens, coke is the best, and should be used more than it is. Hard coal is good and makes a hot fire, but its extra expense is an objection. It is a good plan to mix a little in the fire with the coke, when a very hot fire is wanted. Most ovens are fired with soft slack coal, on account of its cheapness. It is good for drying rough work, but a serious objection is the soot and dirt it makes.* Look at a moulder after he has been inside a cylinder, brushing or cleaning the mould, and he is so black and dirty that you would hardly know him. The question is, can a man in such a condition feel like doing a clean, neat, mechanical job? A soft coal fire is not a steady fire, and it requires close attention to keep it going, and will burn moulds and cores very often. Moulds or cores blacked dry, with this kind of firing, generally make a rough surface, on account of the oily substance the smoke and soot leave on the face. With a coke fire, moulds or cores can be blacked dry in a neat, clean manner, being almost as clean when they come out as when they go into the oven. It makes a steady fire that needs very little watching; and with a fire basket, like the one shown in the cut, you need not bother the night-watch-

* A plan which works well in ovens for using slack coal is illustrated on p. 225, Vol. II.

man to do any firing during the night. Should your mould or core be burnt, you can blame no one but the man who fixed the fire before going home.

If you want a good hot fire, fill up your basket full of medium-sized coke, and leave the drafts open.

Should you want a slow fire the first part of the night, so as not to blister the green blacking, or crack open the cores, leave the drafts partially closed, and have the watchman open them in an hour or two.

Should you want a slow fire all night, only have the basket half filled, and keep the drafts all closed. When the fire is renewed in the morning, shake up the grate bars, and run a bar between the upright bars, to loosen up the fire and get the clinkers out, and then put on more coke, and your fire will run all right again till night. A basket of the dimensions shown is large enough to heat an oven 10 feet wide, 18 feet long, and 8 feet high. For very large ovens, it would be better to have two baskets, one on each side of the oven, and for all-sized ovens to have extensions built, as shown attached to the revolving oven (in article "Ovens for Drying Small Cores"), and to have the fire basket placed in the extension. When the fire-places are built inside the oven, there is a large space to be heated that cannot be used to any advantage, which causes a loss of fuel.

In arranging for the fire basket, two bearing bars are built into the brick walls to support the grate bars, and the bottom is bedded on a solid brick foundation, the front of which is left open, so as to admit a draft and to get out the ashes. The back and sides are closed up, so no cold air can get into the oven. What air does get in is heated, as it has to pass through the fire, making an increased draft and combustion.

The top basket frame is supported by the four corner uprights, which have projecting pieces cast on them for this purpose. Should any of the upright bars get burnt out, they can be taken out and new ones put in.

The inside, which is subjected to the direct heat from the fire, should be built of fire-brick, and the whole brick-work of the extension should be well stayed with binders and bolts, so as to keep the heat from cracking it.

This plan of a coke basket and fire-place is from one I constructed not long ago, and it gives perfect satisfaction. Attached to the fire basket is a fire front not shown. The lug, *B*, is one of four that are cast on the face of the basket frames for bolting the front to. The purpose is to make the fire-place all air-tight above the grate bars. The front has two doors that can be opened for shoveling in coke. Also slides open opposite each space in the basket, so as to get at the fire. The slide or damper is made to close the openings, so that, if the thin sheet-iron outside doors, *X X*, were opened, you could see no fire. These doors do not come down within 2″ or 3″ of the bottom, the space being to admit air. When the draft is to be closed air-tight, a loose plate is used to close the bottom opening, and sand shoveled against the joints.

The most essential point in constructing ovens is to have good draft arrangements. In most ovens there should be a top and bottom flue opening into the chimney, with dampers to open and close them. It is a good thing to have a cover placed over the top of the chimney hole, and made to open and shut; and, when you want to retain the heat, and keep the oven from getting cold, shut down the cover, and it will save two or more hours' firing when the oven is not wanted for a day or so.

The styles of fire-places used are various, but the two shown will give good satisfaction. The one above described is the best for drying a fine class of work that requires an even, steady fire; but where you want a strong heat for moulds that will

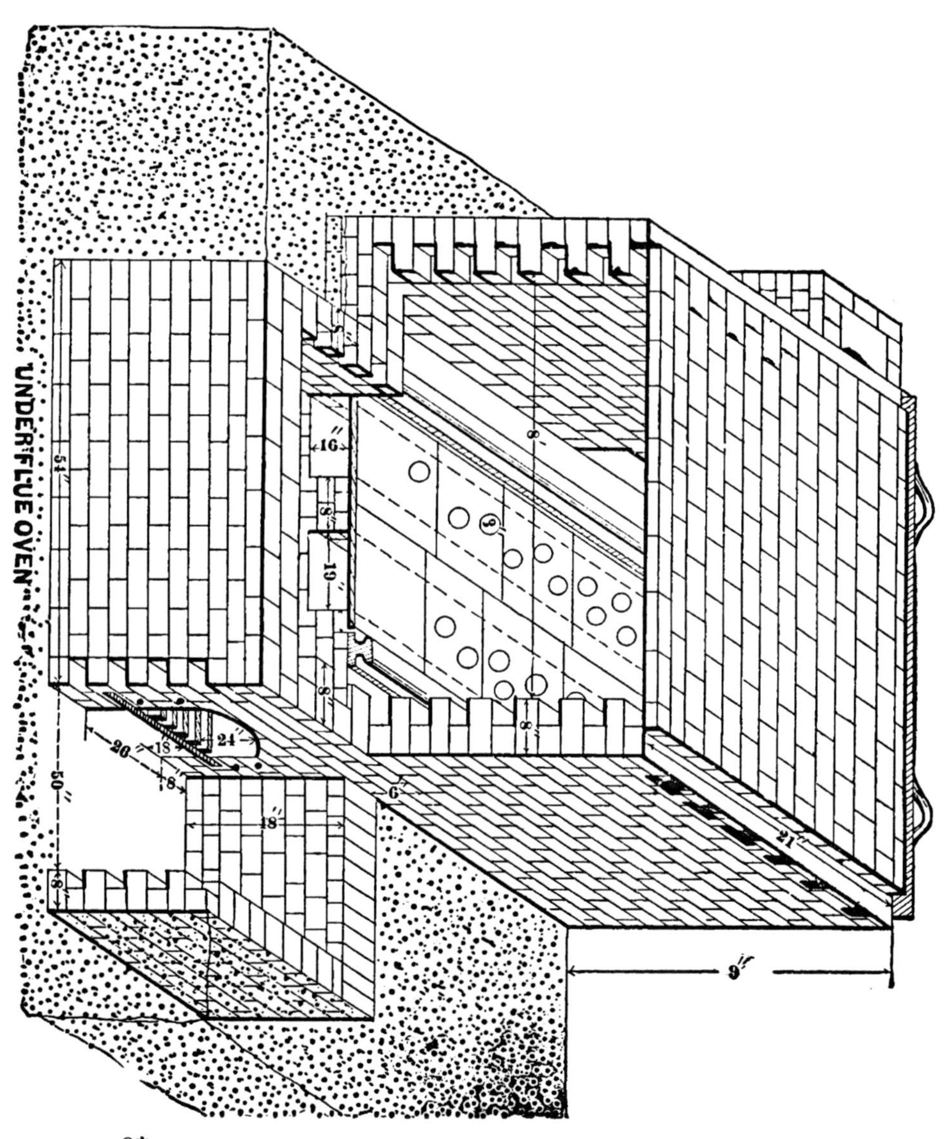
UNDERFLUE OVEN
54"
16"
19"
24"
20"
18
50"
18"
6"
21"
9"

stand to be dried quickly, the under flue firing oven, as shown, will do the best work. In this style of a fire-place the heat comes up from the bottom through the holes made in cast-iron plates. The plates are made $1\frac{1}{4}''$ thick, and in sections, so that they can be lifted by hand to clean out the flues. The oven is shown having the rear end open; also one side of the firing pit, and the arched brick covering over the fire-place, by which means the construction of the two flues that run the whole lengh of the oven can be seen.

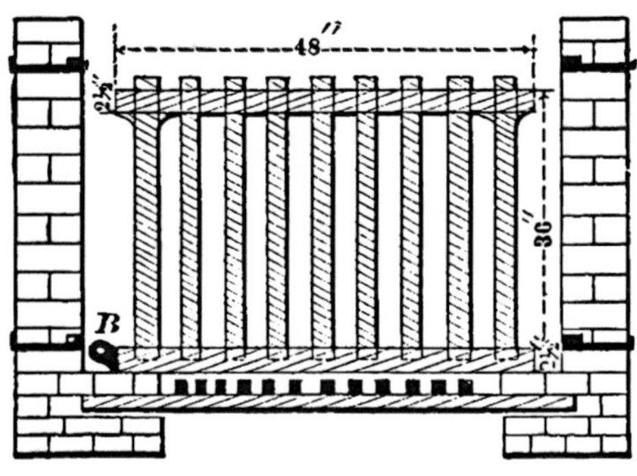

ELEVATION OF A COKE FIRE BASKET

These flues should be built with fire-brick, and it is best to have about six feet of the covering of the end connected with the fire-place arched over with fire-bricks instead of the perforated iron plates, so as to prevent the direct flame from the fire from getting into the oven, or burning out the plates. I have seen strong fires make these plates, at the farthest end from the fire, nearly red-hot.

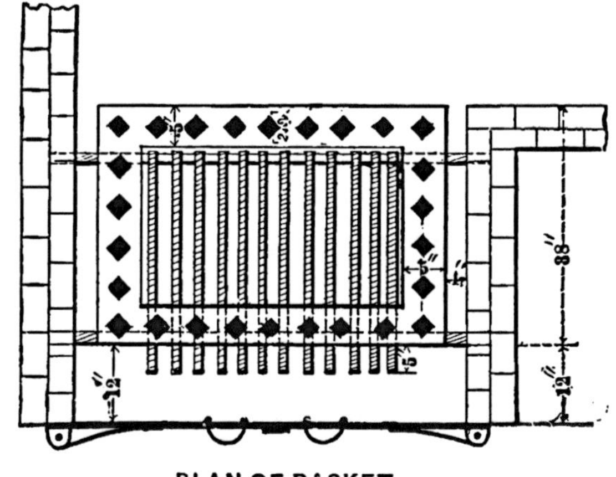

PLAN OF BASKET

The cost of firing such an oven is expensive, as one will burn nearly half a ton of soft lump coal in one night. The fire-place is built below the level of the shop floor, and steps are needed to get down to attend to the fire. Should this pit be built outside the shop, it should be constructed so as to keep the rain or water from getting into the pit.

The size of this fire-place is enough to heat a larger oven than the one shown. This oven shows a flat top, supported by railroad bars, with sheet-iron plates on top of them, and a course of bricks over the whole, to keep the heat from escaping. This style of covering I prefer to an arched brick top, with which there is a large space to be heated that can hardly ever be used.

This flue oven would work well with a coke basket, instead of the coal fire-place, but would not make so hot an oven, or dry as quickly. Ovens could be made having a basket coke fire, as shown, attached to the revolving core oven, and the under-flue arrangement also with a coke fire. To make a very hot oven, both fires could be used, and for slow drying, either one or the other. With this combination, any class of work can be properly dried.

Ovens are now heated by means of steam and hot air, and as illustrating one plan for utilizing hot air we give the following, taken from *The Foundry*, January, 1896, as a description of an oven used by the American Glucose Works, Philadelphia, Pa. "The system we have used and recommend for heating core ovens is the same we have used in drying starch in 40-ton lots (although a little more heat can be used in drying cores) and is as follows: Place to fire outside the oven, but close and alongside of it. Arch over the fire with iron pipes or retorts, so arranged that any or all of them can be replaced. On the end of these pipes farthest from the oven attach a small high-speed blower which can be placed overhead, out of the way. Now run the opposite end of the pipe into a brick flue built inside the oven against the walls on the floor and running all round the oven except across the door. Make this flue at least three times the area of the opening at the fan, and leave a brick out of the flue at intervals all round the oven and on the floor line. These openings should equal the area of the flue.

"Arrange the oven with a door in the top, and which may be opened more or less (according to the number of cores in the oven) to allow the steam to escape. By this plan the fan forces cold air through the pipes where it is dried and heated as hot as required, passing into the oven dry and hot; and with currents of air all over the oven rising to the top, carrying all of the steam with it, the cores will become thoroughly dry and strong in a much shorter time than can possibly be done by any other arrangement. By this plan it will be seen that as all the hot air is discharged into the oven all over the floor, and receives no additional heat after entering the oven, all parts of the oven will be heated uniformly within one or two degrees."

OVENS FOR DRYING SMALL CORES.

It is very important to have in a foundry an oven that will dry cores at short notice, and without burning them. I have seen a large variety of core ovens, and very few of them were good for anything, as they would burn the cores, or require a long time to dry them.

There are shops that have nothing but large ovens for drying small cores, which is all well enough as long as they have cores to fill the oven with ; but to fire up a large oven to dry a few small cores, which is often done, is a very expensive and a slow process.

In building ovens, the builder or designer sometimes seems to have thought all that was required was a fire-place and a space to pile cores in ; whereas, with a little more thought, they might have had an oven that would have been a success, and have cost no more than the apology for one.

The revolving oven, shown in sectional elevation (first illustration), is round, with an upright cast-iron shaft, having five flanges on which to bolt plates or arms *X, X*, the shape of which is shown at *B*. This oven is built with an 8″ brick wall to form the outside, and a cast-iron plate for the top, on which is a box, *D*, to which a cap can be bolted, to hold the top of the shaft, the bottom of which rests in a cast-iron seat.

The fire-place should be built outside of the circle, as shown, so that the cores will not get the direct heat from the fire. In building the walls, hinges *H, H*, should be built in for hanging the oven-door to. This door should be made

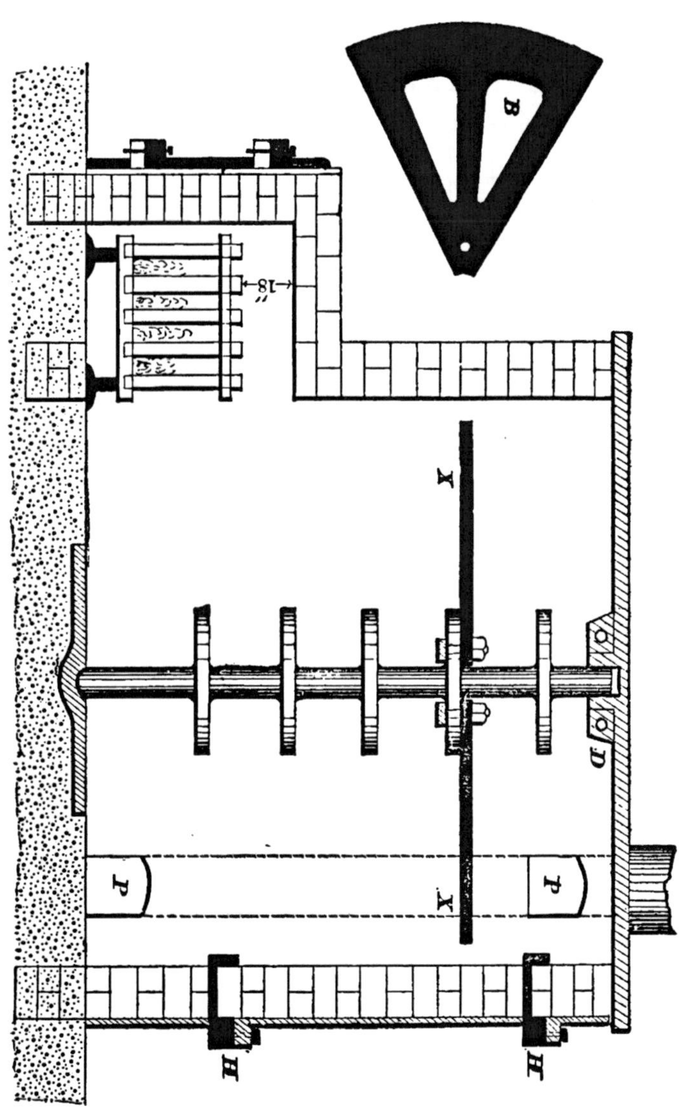
B
18"
X
X
D
P
P
H
H

in two pieces, so as to open to the right and left, and should be the full height of oven, to provide for putting cores on the top shelves. The chimney should have a top flue, as well as a bottom one, as shown at *P P*, and dampers in both, so as to throw the heat down or up, as required. When starting a fire both dampers should be open, and when the cores to be dried are on the top shelf, the bottom damper may be closed, and *vice versa*.

This style of oven is very handy for drying cores that can be lifted by hand, and will hold and dry more cores with less fuel than any oven I know of. Should you want to dry a single core quick, put it on the top shelf, and turn it around to the fire. This oven can be filled with cores, and they can be taken out again without going farther than the door, which alone is of great value to a core-maker. The size of this oven was about eight feet in diameter, and seven feet high.

The second cut is a plan of an oven I constructed, and, for a small one, I think it will be hard to beat. It will dry cores on the bottom as well as at the top, and in a very short time, and without burning them. The amount of fuel used is small. It is made with two doors to open right and left, so that cores can be put in and taken out handily. The loose bars, *X*, can be taken out or moved, so as to make room for large cores. The opening to the fire is at the back, so as to keep all dirt and ashes out of the core-room, and the heat is drawn under to the other end of the oven, and escapes at the bottom flue, the top flue, or damper, being only opened when the fire is started, so as to let the smoke out, and keep the oven clean from soot.

When starting a fire, the bottom damper being closed gives a direct draft, as shown by the arrow. The top damper is made so as to close up the front, but when pulled back it only partly closes up the chimney. The bottom damper is

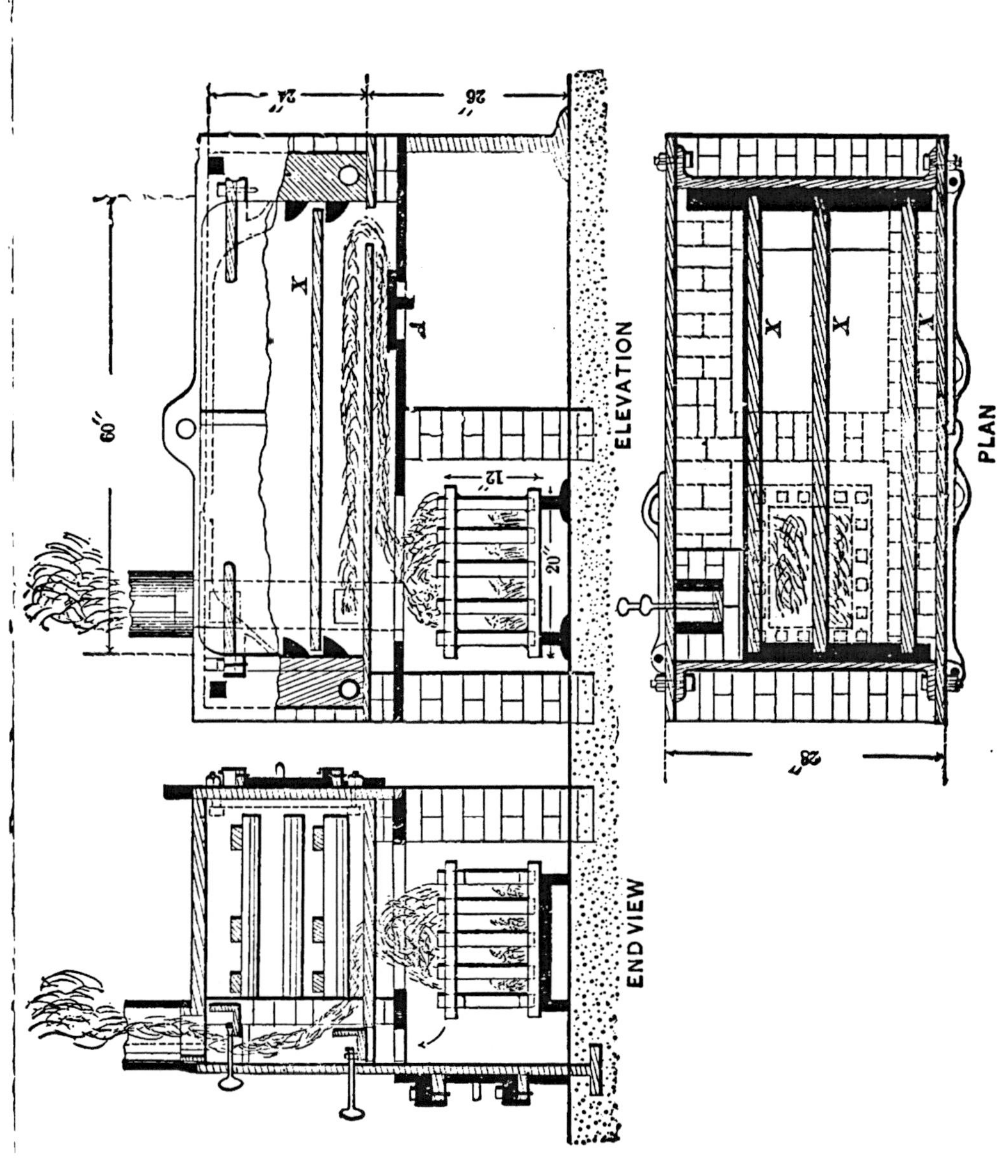
ELEVATION
PLAN
END VIEW
60"
24"
26"
12"
20"
28"
X
X
X
Y
P

made so as to close up the direct draft when the heat is wanted to go into the oven, as shown.

This oven consists mostly of cast-iron plates, brick being used only for the fire-place, between the two bottom plates, for the ends of the oven, and for the chimney, which is at one corner of the oven. The bottom plate, or the one over the fire, should be made about one inch thick, to stand the heat.

The hole and cover shown at *A* is used for taking out the soot, or what dirt may gather between the two plates. For firing, coke is used, which makes a good, clean, and cheap fire, and does not make the surface of the cores oily, as slack or soft coal does.

It is hard to blacken cores dried with soft coal, costing time of the core-maker, and often causing rough spots in the castings.

Since the manufacture of gas from crude oil, etc., has become general, it is economically applied for firing boilers; hence some founders have adopted this method for heating ovens to dry radiators and similar small cores. One plan of utilizing these gases is to build ovens so as to pass the flame into a large open chamber adjoining and below the level of the oven floor, from which the heated gases and air pass along through flues under the oven so as to deliver their heat through checkered brickwork evenly into the oven. This plan for drying small cores in large numbers has been found to be very economical, as it dries them evenly.

TWO WAYS OF MOULDING CROOKED PIPES IN LOAM.

THE adage, "There is nothing new under the sun," must have been written more for the consolation of those who never tried to find anything new, than for men who have spent years or a lifetime in bringing to light some principle or invention. Literally, the adage is true enough ; but, in any event, there is one thing men should get credit for, and that is, the improvements they make on old tricks. The different ways there are of doing the same thing, can only be accounted for by men's minds traveling through different channels to find them. There are all kinds of old tricks or plans for every existing occupation, and also for many that are to come into existence, and in this the moulder seems to have been provided for as well as other tradesmen. If he will only hunt for them, he will be astonished at the number stored up for him.

The cuts here shown represent the result of two moulders searching for a rigging to mould or sweep up crooked pipes in loam. The moulding of these pipes shows the diversity of minds, and the different ways different men will adopt to do the same job, or the same class of work. Each plan here shown has its special advantage as well as its disadvantages, which the practical man can readily see. What would be an objectionable feature for this job, might be a very acceptable one for some other job.

The upper cut shows the process of sweeping the bottom and top part of the pipe separately ; also how the top or cope is rolled over, so as to be closed on the bottom part.

The lower cut illustrates a plan in which the cope was built so as to save rolling over, which, for very large pipes, is a point of considerable importance.

In moulding the upper pipe, there are some principles involved and adopted, in order to have a good chance to finish the cope, and to save labor in the long run. The handles 1, 2, and 3 belong to the bottom plate. No. 4 is a section, showing a part of the cross bars that are bolted to the top plates, between which bricks are wedged, so that they cannot fall out when the cope is rolled over. *E E* show the end view of plates having prickers on. These plates, when bolted down, as shown, assist the holding in of the bricks under them, and the prickers can be daubed up with loam or a dry sand mixture to form the top joint. *H* shows how the two end cross bars are made, having prickers cast on them to hold the sand that is rammed between them for forming the flanges and end joints.

For moulding or sweeping this pipe, a wooden frame, *X X*, having screwed to it the flanges, as seen, is used. This frame is first blocked up level in the position wanted, and then as the brick-work is built up, or the loam rubbed on, a sweep, *D*, the front view of which is shown at *A*, is worked along on the frame *X X*, so as to sweep up the mould.

In making this mould there were no core prints, or bearings at the ends, for the core to rest on. The mould ended on the outside of the flanges, and to form the face of the flanges and core prints there were half-pricked plates used, as shown at *F* on the plan and end elevation. The trunnions cast on these plates made them easy to handle.

When the bottom part of mould was set down to be got ready to cast, one of these half-plates was set up against each end of the flanges or mould, and wedged up until the half-round

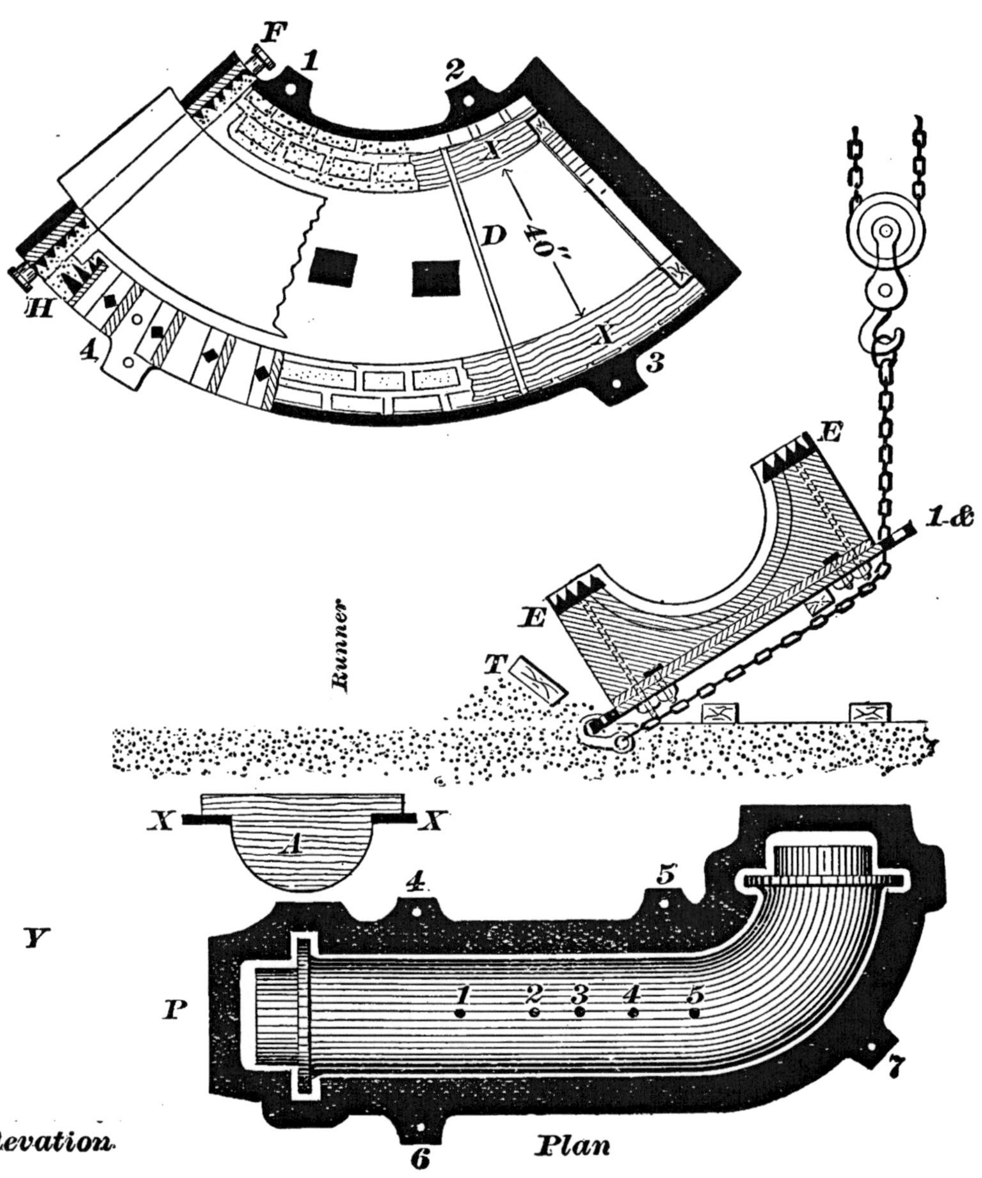
F
1
2
D
40"
H
4
3
E
1-&
E
T
Runner
X
A
X
4
5
Y
P
1
2
3
4
5
7
6
Elevation
Plan

hole in the two measured correctly to answer the purpose of a core print, and after the two chaplets were placed, the core was set in, and the cope rolled over by using two cranes; one crane being hitched to the handles 3 and 4, and hoisted up, as shown, until it came on to the sand pile *T*, so as to prevent any sudden over-balancing. After this, the second crane is hitched to the upper handles, 1 and 2, and the cope hoisted up clear from the ground. Then by hoisting upon the crane, as shown, the cope is completely turned over, after which one crane can handle it, and place it wherever wanted.

After the cope is placed truly on top of the bottom part of the mould, in doing which there can be no trouble, as the top ends are all open so that the moulder can put in his hands and feel the two joints as well as see them, two upper end plates, which are not shown, are lowered over the ends of the core and pressed up against the face of the upper flange, being held there by props, clamps, or bolts. The mould is then ready to be rammed up and got ready for casting; it being poured with two runners at the bottom, as shown.

The lower cut shows a plan in which there is less rigging used, at the expense of extra labor required to make the mould. In moulding this pipe, a frame was used, the same as described above, with the exception of core prints being fastened to the flanges. When the mould was swept up the core prints were also formed.

A, shows the sweep for forming the bottom part of the mould, and *W*, the sweep for forming the outside diameter over which the cope is built. The elevation shows the bottom half of the mould all ready for having the bricks built to form the cope. *P P*, is the iron lifting-plate, which, as seen in the plan, is one continuous plate entirely around the mould, being kept back from the ends of the core prints to allow plenty of room for the core when it is placed in the mould.

Y, shows a reliable way of forming a false mould or case for building the cope over. Sometimes the top, or false, mould is formed of all sand instead of as shown. After the brick-work of the cope is completed, it is best to cover it over with a plate to protect the bricks. If necessary, this plate can be bolted down to the joint lifting-plate. When all is ready, the cope is hoisted off by 4, 5, 6, and 7, and the cope finished overhead. The false mould *Y*, and loose sand in the bottom is then all removed, and the bottom part of the mould finished.

The cores for both of these pipes were made in the ordinary manner; cast-iron core arbors and plates being cast the shape of the pipes on which the cores were swept up with common core sand. When dried, the halves were pasted together for the lower casting, but for the upper one, the halves were not pasted together.

After the joints were rubbed together, so as to make the core of the right diameter, the bottom half of the core was set in the mould, and the top half was set on, without using any paste whatever, thereby saving the labor of bolting the halves together, or using heavy straps to lift the core by.

The figures on the lower cut, 1, 2, 3, 4, and 5, show the position of the pouring gates. This casting was poured by having the iron drop on the top of the core, which, for thin castings, is better than having the casting poured from the bottom, in which case, by the time the iron fills up to the top, it is apt to be dull, and cause the top of the casting to be cold shut.

Sometimes, when making such pipes, the core is made similar to that known as a loam core. The bottom half of the core is made by using an iron pricked frame the shape of the pipe wanted. The frame is set on an iron plate made also the shape of the pipe, such plate being 1½" wider than the diameter of the core, so as to leave room for the sweep to work in. When forming the core, use the prickered frame as the center portion, filled with coarse cinders or gravel, to take the vent off; the frame is then wedged full of bricks, so as to leave room for an ordinary coat of loam which is swept on, and the bottom half of the core is formed by the use of such a sweep as is shown at *W*. This half core is run in an oven and dried, while the bottom part of the mould is being swept or bricked up, as shown by the elevation of the lower plan of making a pipe. When the bottom half of the mould is ready, the half core is taken out, and rolled over on a bed of sand; then, by the lifting-hooks or screw holes provided in the iron core frame, it is hoisted up and set in the bottom half of the mould, the space left open to form the thickness of the casting is then filled up with dry sand. This leaves the bottom half of the mould formed as far as the making of the core and mould is concerned. The frame which was used for sweeping up the bottom part of the mould is reset, and the top half of the core is bricked and swept up. A thickness is then swept on over the top half of the core, and after the wooden flange patterns have been secured to their places, the forming-frame is then taken off, and a joint lifting-plate, as shown at *P P*, is then set on, and the cope built up. After being lifted off, the thickness is taken off, and the whole core is hoisted out, and each part being finished, is separately set in the oven to be dried. The closing of the mould, and chappleting of the core, is done in the same way as if a dry sand core had been used.

Square pipes are often made after the above plan, as far as principle is concerned; the ways and manner of sweeping are often changed in order to form different angles.

To make the thickness on a straight side, where it would be difficult for green sand to stay, often a flat loam cake is nailed instead of waiting for loam and pieces of bricks to get stiff. In lifting a cope from a square print or any part of a mould, where there would be danger of it breaking or pulling up the mould or prints. It is a good thing to lay some thin slabs of wood between the two parts, and before the loam gets too stiff or hard pull them out, and thus leave an open space between the core and the outside mould, allowing a little play when lifting off the cope.

MOULDING LARGE QUARTER-TURN PIPES IN LOAM.

THE versatility of loam moulding, or the aptness of the moulder to change from one course of treatment to another, is generally caused by some crookedness in the shape or form of the casting to be made.

The quarter-turn pipe pattern shown is a full wooden one, and the moulder who had it made must have well considered all the essential points before ordering it, as the cost of making such a pattern must have been considerable. As this job was a standard one, the full pattern would pay for itself in the saving of labor in moulding it in a short time.

Upon page 159 is advocated the use of sweeps instead of patterns for making loam castings, and lest some may think there is a lack of harmony, they should remember that I advocated their use when it was practicable to use them, and not under all circumstances.

In the case of this quarter-turn pipe, there is not much more than two feet square, but would require a sweep of a different shape to sweep the mould with. Some will say the pattern might have been made a skeleton, or the pipe cast flatways by using a frame and sweep. I think, however, the plan shown is the most practicable.

These quarter-turn pipes were bolted to the large pipes cast in green sand, the manner of moulding of which is shown in page 78. In making the rigging for moulding the quarter-turn pipes, the bottom plate and lifting ring were cast very thick, as they had to carry a heavy weight. Had they been lighter, they would have been liable to have sprung, so as to cause the mould or brick-work to crack open.

The bottom plate is set solid on iron bearings. The lifting-ring is then set on even and true with the outside edge of the bottom plates, after which the full pattern is set on and blocked up in position as shown. The brick-work is then built under it, using a light straight edge, and having for a guide the edge of the lower flange, and the inside face of the lifting-ring to form the bevel joint, which separates the outside from the inside part of the mould. This joint also forms a guide to close the mould together by.

The inside face of this lifting-ring should be well oiled when it is first set on, to prevent the wet loam from sticking to it. When finishing up this joint, it is well oiled, and parting sand is sprinkled over it. Charcoal blacking, wet with water, could be used instead of oil, to make the outside part from the bevel joint. Charcoal is a very light substance, and when the water evaporates from it, the charcoal returns to its original dusty state. Therefore when the charcoal blacking is brushed on to form a joint, the loam, as it stiffens, absorbs the water, leaving a thin layer of dusty charcoal, which makes a joint between parts of moulds that are to be separated.

When building this brick-work under the pattern, it is built so as to form body, or thickness of wall, enough for the core to be built on; and at the same time the core is built up 2 or 3 courses, or layers of brick, so as to hold the pattern in place, and to form a guide by which to set the pattern back to finish building the core. The outside of the mould is built first, and then the core. To make a joint, so that the outside part of the mould can be made to separate in halves, one half of the outside is first built, and then an upright joint is made, using for a guide the parting in the pattern. This joint is then oiled or blacked, after which the other half of the outside is built up.

Some of these pipes had a branch cast on them, as shown. The small pricked loamed plates, *E, E, E,* which should be about two feet long, are built in with the brick-work for supporting under and over the branch when the pattern is drawn. When the mould is being closed together to cast, after one half is closed on, the round core, *Y*, is set in a round print formed about three inches deep in the main core. The other end of this branch core is bolted back against the half print, as shown. The brick-work is not shown on the side of the pipe. This gives a clear view of the wooden pattern, loam plates, and branch core. When the brick-work is built up nearly to the top, a light cast-iron ring, *D, D,* split in halves, is set on to strengthen the brick-work. The outside is then bricked up to the top, and the top joint made.

The pieces of wood, 1, 2, 3, 4, and 5, that are screwed on the pattern to hold the parts together, are unscrewed and taken off. The three-winged cast-iron cross, shown at *X*, has three chains hooked in the staples, there being two cast in each wing, so as to give a better chance to regulate the lifting off. In the cut there is only one half ring shown. When the outside parts of the mould are ready to be separated, the lifting irons or bolts are hitched in the three ring handles, *B, B, B.* Half of the mould is then hoisted a little, and should it not hang just right, lower it down and adjust the stirrups till it hoists level.*

The half mould, when hoisted off, is pushed around to the oven carriage and lowered on it, which operation is repeated with the other half. The patterns are now drawn, and the moulds finished and run into the oven to dry. The two

* An excellent chapter upon the principle of "balancing and hoisting moulds" is seen on p. 435, Vol. II.

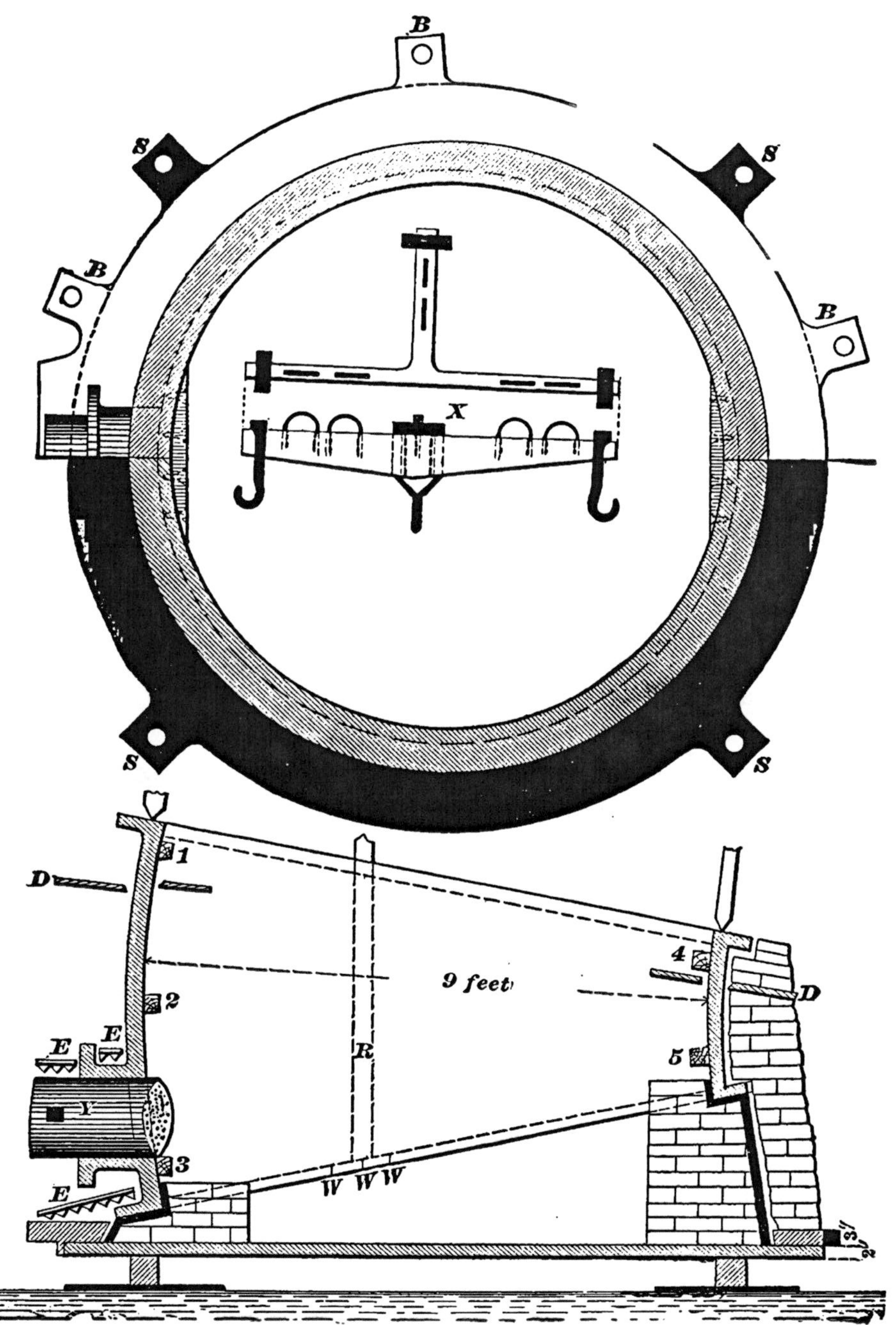

MOULDING QUARTER-TURN PIPES.

half patterns are now set back in their original places, and pieces are screwed to the outside to hold the two parts together. The core is now made, and with a four-winged cross hitched to the four handles, *S, S, S, S,* the bottom plate and core are also hoisted on an oven carriage. The pattern is drawn, and the core finished and run into the oven to dry. While these are drying, a top ring (not shown) is made to cover the top flange with.

When all the moulds are dry, the bottom and core are hoisted off first and set level where wanted. The outside halves are then placed on the bottom in their place, and the top covering ring set on. The whole mould is now ready to be bolted together, which is done by bolting the top covering ring down to the bottom plate, the handle being used to bolt the two together. The mould is not sunk in the floor or pit, but is set up on the shop floor and the sand rammed up around it, a staging being used to pass the sand up.

When the mould is cast, the bolts in the sheet-iron curbing are taken out, and the curbing taken away. Then, by digging around the bottom of the sand, the upper portion will fall down.

W, W, W, show three gates cut into the flange, and *R* the upright runner, one of which is on each half of the mould. A flow-off gate is on the low side of the pipe, and a feeder on the high side, as shown. The castings made in this way were good and solid.

MOULDING KETTLES IN LOAM.

In the engraving is represented the common method of making kettles in loam. The bottom plate *X* rests solid on three or four blocks, as shown at *PP*, the inside sweep is attached to the spindle and the bevel-joint *D*, the top of the flange and the inside of the kettle is bricked and swept up. After the loam has become stiff and hard, the outside sweep is attached, and a thickness is swept up, on which the flange and the outside of the kettle are formed. This thickness is generally formed with green sand, keeping the sweep up, so as to sweep from $\frac{1}{16}''$ to $\frac{1}{8}''$ thicker than the casting required. This gives the required thickness when the sand is sleeked. Over this sleeked surface a coat of clay wash is brushed and dried hard, thus making a solid surface to build against. To form a joint between the thickness and outside, oil and parting sand are used. Sometimes, instead of keeping the sweep up, to allow sleeking, and clay washing over the surface, the sweep is set to give the thickness wanted, and after the green sand thickness is roughly swept up, a thin coat of loam is swept upon the green sand, thereby forming a smooth surface.

There is not much trouble in sweeping up a plain surface thickness on loam moulds with green sand, but where there are flanges or projections it requires time and patience. It is a good plan, for instance, instead of sweeping up the flange thickness with green sand, to form it with pieces of brick and loam. In some instances this plan is adopted.

Sometimes wooden segments are used to form the flange of the kettle. After putting on the thickness the joint is cleaned off and oiled, and parting sand sprinkled over it, after which the outside lifting-ring, *HH*, is set on, and the outside of the mould bricked up, as from *B*. After the thickness has been swept up, the spindle is hoisted out and the hole firmly bricked up. The outside being bricked up, it is then hoisted off by the four handles *H*. The thickness is then taken off and the mould finished up and put into the oven, or dried on the floor.

In getting ready to cast, a sheet-iron curbing is set around the outside of the mould, and sand rammed between it and the brick-work, the same as in similar loam moulds. After this sand has been rammed about six inches above the top of the brick-work, a flat plate is bedded on and wedged or held down by the use of an iron cross, and slings hitched to it and to the four handles of the bottom plate. When ramming this sand, care must be used, as it is not like ramming up a plain vertical loam mould. The pounding of the rammer should be lighter the higher up it is used ; in fact, the upper parts do not require hard ramming.

The lower part of the mould should be rammed solid and hard, as there is considerable strain there ; but for the top, if the sand is firmly tramped and the plate solidly bedded down, it will require but very light ramming.

For running kettles moulded in this way a number of small grates are in general set around the top, one being shown at *E*. If run from the bottom, such castings are likely not to be solid, because the iron gets dull before it reaches the top, and also because the dirt has a better chance to gather in lumps or streaks, thereby making spongy iron. Even when the casting is poured from the top there will be more or less dirt, but it will not be so bad as when run from the bottom.

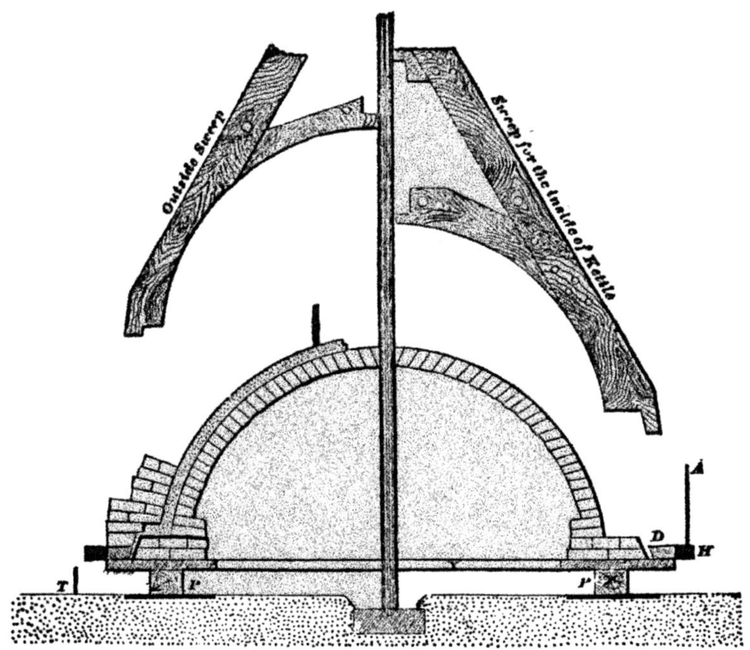

DEVICE FOR MOULDING KETTLES IN LOAM.

It is very important in casting kettles to properly carry off the vent from the inside of the core.

Not many years ago an accident happened to a moulder, an acquaintance of the author, that came near costing him his life and setting fire to the shop, by the blowing up of the mould when being cast. When the mould was nearly full of iron, there was an explosion that threw out the most of the iron in the mould. The trouble was in making no provision for vent except one small tube or pipe, and the mould being poured fast, gas was generated rapidly, producing what is sometimes called fire-damp. There being only one pipe and lighted with shavings, the gas took fire, and running downwards to the gas inside the brick mould, it instantly exploded.

In moulds that have a confined air space, when the gas of the mould is driven by the heat of the melted iron it exerts a pressure. This pressure, if given a good chance to escape, relieves itself without doing any harm. To provide a proper escape, the bottom part of such moulds are better if left open; as, for instance, in ramming up the mould, shown in the cut, instead of letting the curbing come down below the bottom of the mould, as shown at *T*, let the curbing rest on the handles, or on some blocks, so as to be up above the bottom of the mould, as shown at *A*. By this means, when the mould is rammed up, the underneath portion is all left open, and when the mould is being poured, no fears of an explosion from foul gas taking fire need be entertained.

This is applicable to the casting of steam cylinders having one head cast in, or hollow castings that have the bottom cast up.* Many moulders, to avoid trouble with such castings, will fill up all the open space with dry dust or fine cinders.

If a mould cannot be left entirely open around the bottom, there should be two pipes, or openings (the larger the better),

---

* A pit especially designed for this class of work is illustrated on p. 228, Vol. II.

which will form a draft and give more chance for the gas to escape.

When gases explode, the explosion is caused by heat or flame. Gases can be raised to such a temperature as to ignite themselves, which will account for the explosion of moulds where fire cannot reach the vents.

Moulders sometimes use what is called a cold vent, which works well for some classes of green sand moulds ; but for such loam moulds as the one shown in the cut, they are not used.

A cold vent is one where the vent pipe is led away from the mould to insure it against being lighted by any flying sparks, so as to have the vent come off without burning.

If we can by setting fire to shavings underneath a loam mould or core, set on fire the inflammable gases as soon as they are driven out of the mould, the explosion, if any occurs, will be very light, and with the gases once on fire, and good outlets for their escape, there will be no danger. But should we wait until there is a large volume of gases generated and then set fire to them, it will be dangerous.

CASTING ANVIL BLOCKS.

UNDOUBTEDLY the heaviest castings ever made have been for anvil blocks. One casting for this purpose, made in Russia, weighed 600 tons, while in this country one has been made weighing 160 tons.

The cut represents different ways of moulding anvil blocks. The main point to be considered in making such a class of casting is to have good, solid, ground bearings and a strong bottom plate, so as to support and not allow the bottom portion of the mould to sink away when the weight of the heavy mass of iron comes upon it.

The heavier the mass of iron the thicker should be the under brick-work. A body of metal that will keep in a liquid state for two hours should not have less than two layers of bricks to form the bottom of the mould, and for each additional hour there should be added a course of brick, until there are six or seven courses, which should be sufficient for any casting. For very heavy casting the bricks that are used to form the face of the mould should be good, first-class fire-bricks of a soft quality. For the bottom part of the mould it is better to have at least two courses of fire-brick.

It is better to use an iron curb when ramming around the sides of such moulds, than to depend altogether on an earth or brick pit.

For bolting together heavy anvil moulds strong binders should be placed under the bottom plate, as shown at *XX*, and similar binders should be placed over the top of the mould. To bind the mould, having large surface copes, with

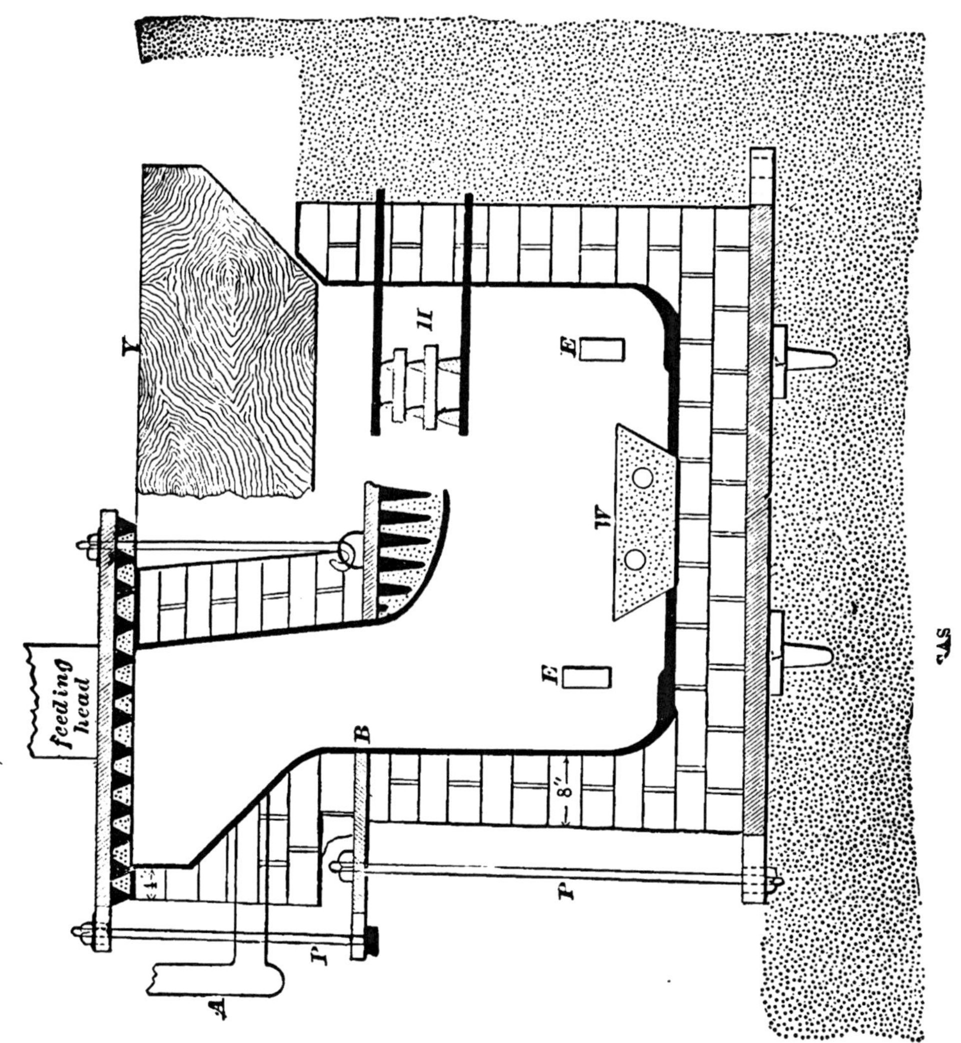
feeding head
A
P
B
P
E
E
W
H
Y
8"

bolts through the handle of the plates, as shown at *PP*, is not a very practical plan, as it gives the center of both plates a chance to spring.

Sometimes there is an iron cross used over the top plate, and, by having the lower plate handles project out far enough, slings or chains are hung down from the cross, and when hitched to the lower plate handles the cross is hoisted up until the slings or chains have a strain on them. Then blocking is wedged between the cross and upper plate, by which means the mould is held together without bolts.

The foundation plates for building such moulds upon are made from 2″ up to 5″ thick.

The casting of very heavy anvil blocks is generally done on the spot where they are to be used, and after they are cooled the block is turned over by means of wrought or cast iron trunnions cast into or on the block.

The common plan of making ordinary anvil blocks is shown at the left-hand side of the cut. An 8″ wall is built up to within about 6″ of where the pattern begins to extend out, and, after a joint is partly formed, a center-plate or ring *B* is bedded on and a parting made; then the balance of the mould is bricked up. The reason for using the center-plate is to save the work and the drying of a thick wall, which must be built if this plate is not used, and also to part the mould, and thus afford a better opportunity to finish the bottom portion.

Another plan is to build only the straight portion of the mould in loam, and when this is dried and rammed up level with the top of the brick-work, then set on the top portion of the wooden pattern, as shown at *Y*; then ram up the balance of the mould with green sand. After the pattern is drawn and the green sand portion of the mould finished, the mould can be covered with a loam plate or a dry sand cope, and if it is not convenient to bolt down the covering it can be held down with weights. This method of moulding is sometimes adapted to save labor and time.

Sometimes anvil blocks are cast open, that is, without having any covering or cope on them, and sometimes there are heavy anvil blocks cast at blast furnaces. The mould will be made in loam in the floor, and there dried, and after it is dried the outside of the mould is rammed up. When all is ready the anvil block is cast, there being no cover or cope over the top of it.

For casting anvil blocks wanted in a hurry, time could be saved by making a loamed plate and having it dried by a fire underneath, then set this plate on a level bed down in the hole. The pattern could be set on this loamed plate, and then the entire height of the anvil block rammed up with green sand; that is, providing the sand is good enough. By this means the bottom would be formed with loam and the sides of green sand, and although the sides might look rough there should be a smooth bottom.

The section of a loam core shown hanging to a section of the loam plate covering illustrates the way a large heavy anvil block was made upon which I worked about twenty years ago. This core was put in for the purpose of giving a large base with little metal. At least I think that was the reason.

Another thing sometimes practiced in the casting of anvil blocks is to make some holes about 1½″ diameter through the brick-work, and before ramming up the mould place in bars of 1¼″ round or square wrought iron, as shown at *H*. Bearing on the lower rods, pig iron is piled and wedged down by the top bars so as to keep the pigs from floating. This pig iron is placed in the center of the mould to assist in the cooling of the hot metal. The advantage of this can, I think, be readily seen. It is best to have the pig iron that is to be placed in the mould thrown into the cupola just before the bottom drops, so as to have all the rust burnt off from it before it is placed in the mould.* In the dove-tailed core, *W*, can be seen two holes. In a corresponding section of the sides of the mould there are also holes made larger than those in the core. When the core is set into its print and placed right, rods are passed through the core from one side of the mould to the other, and then the rods are wedged down by filling up the holes in the mould or brick-work with pieces of brick and stiff loam.

This is a safe plan for holding down the core, and is far better than driving nails into the mould at the ends of the core. *EE* shows the bottom pouring gates, through which the mould should be more than half filled before any iron is allowed to go in through the top pouring gates *A*.

The faster such moulds are filled with iron the better, and the iron should be on the dull side. The duller they can be poured the smoother the casting will be.

The patterns for such castings should be split, or made in two sections, the piece *Y* being the top section. By being so divided it will give the moulder a better chance to make his casting.

A nice smooth skin on such massive castings is very seldom obtained, and some moulders will say that it cannot be done because of the large body of iron staying in a liquid state so long. It is, however, chiefly because the loam mixture and blacking is not as it should be, rather than because of the heavy body of liquid iron.

---

* Good judgment must be used in placing the pig iron in a mould, as too much and its position could easily cause a casting to crack.

SWEEPING AN OCTAGONAL LOAM MOULD.

THE sweeping of loam moulds that are not cylindrical or round in form calls for an entire change in the manner of operating, and often the moulder's skill is tested in trying to invent some rigging that will work well. There are two or three ways to mould any casting, and that plan should be adopted that will cause the least risk of losing the casting. The plan adopted by one moulder might not be used by another, each one seeing different ways of handling the job so as to insure its safety or to do it quickly. When speed and safety are considered and combined, it will sometimes require the highest mechanical ability and judgment to make them work together successfully, and to tell whether the moulder has adopted the best plan for moulding an intricate piece requires the judgment of a thoroughly practical moulder.

It is a question if there is not more pattern-making done for loam moulding than is necessary. This statement is not made to deprive the pattern-maker of work, but to intimate that the less pattern-making there is done for loam work the better it is for the mould and casting. When there is a pattern used, and it is drawn out of the mould, the surface of the mould presents an uneven face, full of hollows. This is caused by bricks being laid on top of others, under which the mud is not yet dry. If the loam between the pattern and brick is set, and cemented to the brick, when the top brick is rubbed and laid on, it sometimes presses the under one back, and, the loam sticking to it, leaves the hollows as stated. In some cases the loam, not being set, will stick to the pattern, and when the brick is pressed back there will be sometimes a cavity left between the loam and brick, and when the pressure of the melted iron comes on it, it may press it back so as to cause a swell on the casting; or, should the air or gas confined in this cavity not find escape through the joints or brick, it will pass through the loam, or face of the mould, into the liquid iron, and cause a scab on the casting. The only remedy for this is for the moulder to use the mud as stiff as the job will allow, and lay on the bricks with care, and be sure they are properly pressed up.

When working without a pattern this care need not be taken, nor time lost. Allowing there are no risks from the above causes, there are still other objections to the whole pattern, one of which is the extra work involved in finishing a mould with loam built against wood. The wood must be rubbed over with oil to keep the loam from sticking to it, and with the best of oil some of the loam will stick to the pattern, causing the face of the mould to be started and leaving thin flakes of loam hanging to the surface of the mould. Or perhaps the pattern, although well soaked in water before using, will expand so that some joints or portions of it will project beyond the general surface; and when the pattern is drawn, it will start and pull down the loam. In finishing such moulds, if the oil is not all washed from the face of the mould, the loam put on to fill up the hollows will not unite well, and will be liable to scab the casting, or the blacking will not hold fast to the mould when it is poured, thereby causing blacking scabs, which are very aggravating to the eye, and are so thin that, if chipped off, the white spots will look worse than the scab, if left on, and what to do with them is sometimes quite annoying. When loam is rubbed on to the face of bricks, and then swept off with a revolving sweep, or with a sweep worked by hand in any direction, as may be required for sweeping between frames or skeletons of patterns, the moulder is sure that his loam is packed on to the surface of the bricks in a reliable manner, and with the sweep and fine loam he can make the face of the mould so smooth as to require little or no finish ing with tools, which saves labor ; and not only that, but too much sleeking with tools is ofttimes a cause of scabs on castings.

The octagonal cuts shown are the top and bottom view of a steel ingot mould, the elevation of which shows the length and thickness. The engravings show also the way it was moulded.

Cast iron ingot moulds for standard steel ingots, such as are used in the manufacture of steel rails, are sometimes made in dry and sometimes in green sand, and the rigging for moulding them is constructed so as to make them rapidly. Ingot moulds weighing about 2,200 pounds, a moulder and a helper will make four or five of in one day. The ingot mould shown in engraving took over one week to make. It was made to cast a steel ingot for a large shaft, and may never be used again. It was made with the least expense of pattern-making and rigging possible, and seems appropriate for illustrating sweeping that cannot be done by the ordinary means ; also for showing a plan of gating and running that can be used for other castings with good advantage. This casting being over nine feet long, and smaller at the top than at the bottom, it would not answer to drop the metal from the top, as in falling it would strike and cut the slanting surface of the core, and cause the casting to be scabby, which would condemn it. The inside of these castings must be smoother and more regular than most other castings I know of. If a cylinder has a clean scab inside, it can be bored out, but if an ingot mould has a scab, or even a swell, on the inside of it, it is taken to the drop to be broken up. After the steel poured into an ingot mould is set, the mould is hoisted off the ingot by two staples, one of which is shown at *P*, and should there be a swell or scab on the inside surface of the mould, it would prevent the steel ingot from coming out of the mould. Should the scabs or swells be chipped off, the broken skin of the iron would allow the hot steel to eat into it and unite the steel and iron.

The length of this casting being so great, it would not be safe to have run it all from the bottom gates, as the iron would be dull before it reached the top, causing the casting to be cold shut, which would also condemn it. When moulding it, we made a runner a little above the middle, as shown, so that when the metal running into the mould at the bottom runner came up to the top runner, we could see it by looking down the large riser *D*; at which point we lifted the iron plug *W*, and the hot, clean iron then ran into the mould through the top runner. Should this iron have run into the mould before the iron from the lower runner reached this point, the top iron would have run against the face of the mould, and probably have cut or scabbed the mould; whereas the top iron, not being let in until the bottom was run up, prevented the top runner from forcing the metal against the face of the core, and, as it was iron running into iron, the danger of cutting the mould was very slight.*

The cut *R* shows a plan of the pouring basin, the distribution of the lower and upper pouring gates, and also the riser. The runners and gates were all made in cores, and set one on top of the other as the mould was rammed up. The lower gate and runner were made larger than the upper ones, on account of more iron having to be run in and be forced up to the top.

* The two runners were fed from one basin, because there was only one crane which could reach the mould.

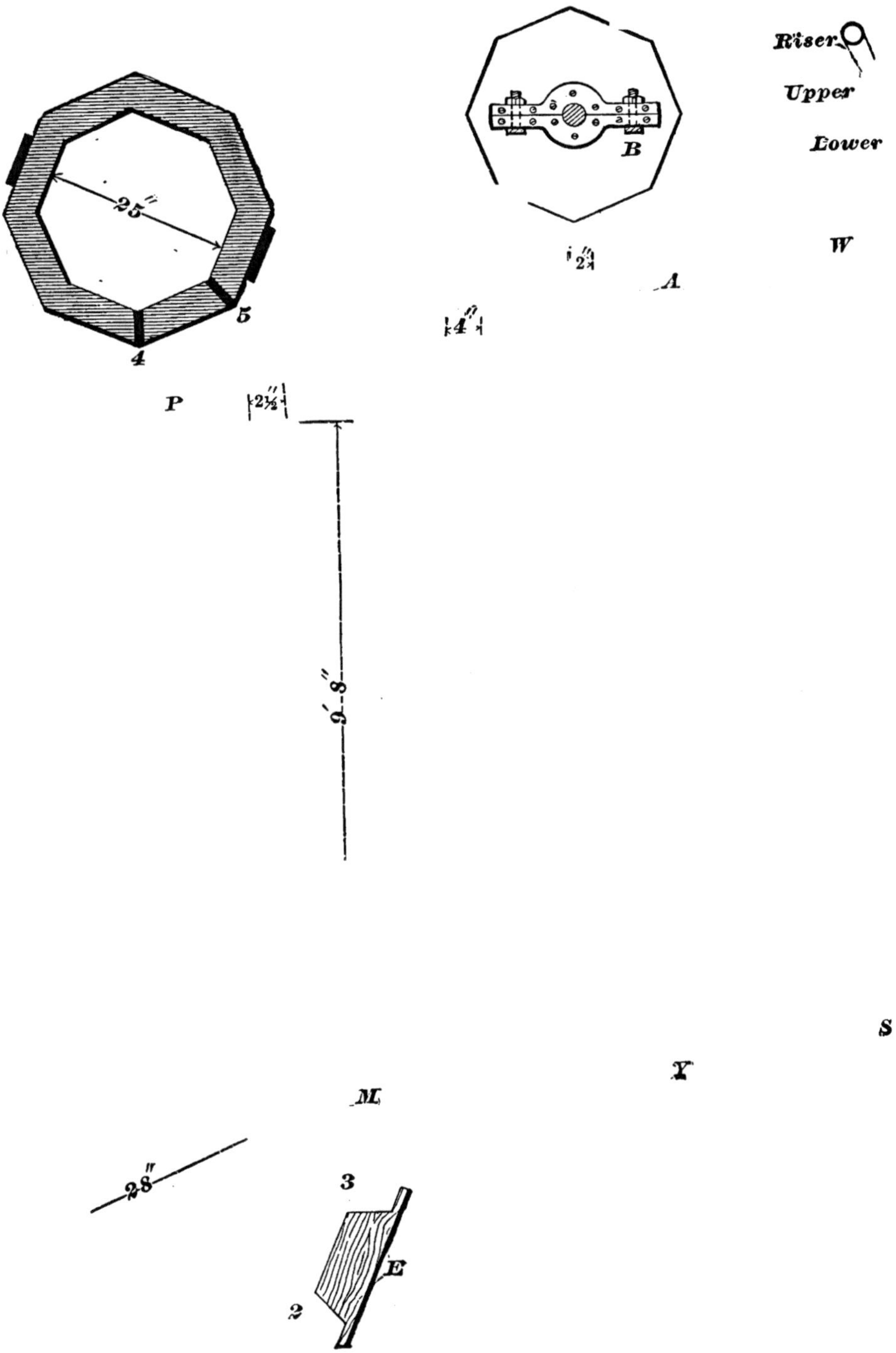

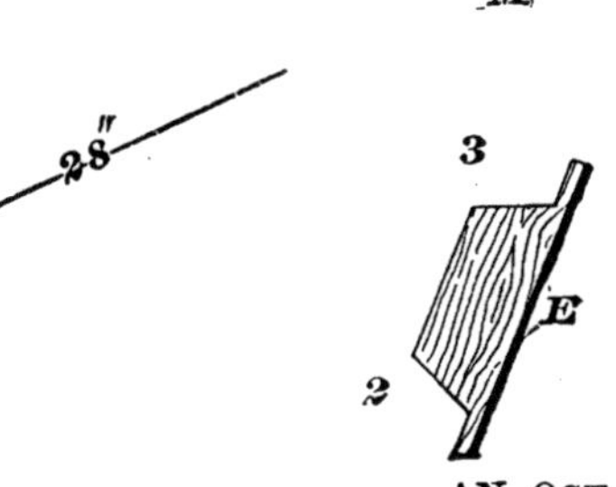

AN OCTAGONAL LOAM MOULD.

When a casting is poured from the bottom it requires a larger runner than if poured from the top, as the more a mould fills up the slower the iron goes in. The iron may rise very fast on the start, and, before a high mould is filled to the top, go in so slowly as to cause the casting to be cold shut.

When moulding this ingot casting, an octagonal frame, *X*, *X*, the same size and form as the bottom of the casting, was used as a guide in building the foundation and beveling joint for the outside to be guided off and on by, and when the bricks were laid high enough to admit of placing a runner under the mould, as shown at *M*, *M*, a sweep was fastened on the spindle, and a level bed of loam was swept up, and the frame *X*, *X*, laid on it and centered. After this the outside lifting ring *S* was put on, bricks built up level with the frame *X*, *X*, and the wooden plate *B* bolted on the spindle, about 8″ above the height the mould was to be made. To get the faces of the top and bottom frames parallel with each other, a plumb bob was hung from one of the top corners, reaching down to a corner on the bottom frame, and, when set right, the top was bolted tight to the spindle, and the outside bricks were laid. About every two feet loam was rubbed on and swept off with the long strike, or wooden straight-edge *V*, the bottom of which has for a guide the inside of the frame *X*, *X*. After the first two feet were built and loamed up, a light, handy straight-edge, put against the face of the mould, was used for a guide in laying the bricks. After the outside was built and hoisted off, the spindle was set back, and the core was built in a solid, reliable manner. Had this core been a round one, an 8″ wall would have been strong enough to resist the pressure for the lower two feet, and the rest of the way a 4″ wall would have held it. The pressure of iron on a round core is the same as the pressure on a stone or brick arch.

If it is built right, the more pressure the greater the resistance. But to attempt to build a flat surface of stone or brick alone to support or withstand a pressure, either in moulding, bridge or house building, would be the height of folly.

On account of this mould being too high to be admitted into the oven, it was parted as shown. The outside was parted lower down than the core, so as to give a chance to daub up and dry the joint when closing the mould; for, should there be any unevenness, or a fin at this joint, it would condemn the casting. The top section of the core is lifted by two hooks *H*, and the bottom section with the foundation plate. The arm *A* is for holding the top of the spindle steady, and *Y* is an iron block bedded in the floor, and having a seat to hold and center the bottom of the spindle.

The cope or covering used was a perforated iron plate, daubed or rammed with core sand, and having the lifting hooks or staples *P* built in it.

This ingot casting could be moulded by having upright strips of wood fastened into the top and bottom frames, as shown at 2, 3, 4, and 5, thus making a skeleton pattern with which to build up the center core first, by using the hand sweep *E* between the upright frames, for a guide in laying the bricks and putting on the coarse loam. To finish the core with fine loam a long straight-edge should be used, on account of the casting being thicker at the bottom than at the top. When the core is loam finished, fill up between the upright frames with damp moulding sand, in a solid manner, and sweep it off even with the outside, over which brush some charcoal wet with water, or oil, with parting sand sprinkled over it to make the outside part form the core. When all is ready, build up the outside against the thickness surface, and with the center core bolted down to the bottom plate, hoist off the outside.

Should the question be asked which is the best plan by which to make a good casting in the least time, I should answer that the plan fully shown with cuts is the one I choose after considering all the essential points.

BUILDING OR LAYING BRICKS FOR LOAM MOULDS.

THE proper laying of bricks is as important a process to the loam moulder as it is to the mason, since they form a support and outline for the inner and smoother and ornamental part of his work, and to build up brick walls or cores so as to stand the pressure of the iron when poured into a mould, and also to hold together moulds and prevent them from cracking open when moved or hoisted with the crane, is a feature in laying bricks that a moulder must be particular to do well. I have often seen loam moulds crack open, from no other cause than the failure of the moulder to break joints when building up his brick-work. The cuts *Y* and *S* show the way brick-work looks when carefully built and when carelessly built. *S* shows all of the joints broken, or the bricks laid as they should be, while *Y* shows the reverse. Some might say they could not break joints because they had not enough whole bricks to work with. This in some cases may be true enough ; but is it not also true that there are many bricks broken unnecessarily ? Some foremen will allow helpers, when stripping off a casting, to take pickaxes and sledge hammers, and knock down the bricks in such a careless manner that hardly a whole brick will remain from a mould. Some moulders, when requiring half or a piece of brick, will break whole bricks, to save the labor of stooping down and picking up pieces. In any pile of brick that have been used once, there are plenty of sizes and forms to be found without breaking up whole ones to make them.

The time lost looking for every little piece of brick might be urged, and of course there is time lost, and the mould constructing may be delayed by stopping to look for pieces; but in building the next moulds out of the same pile of bricks it will not take the moulder or helper so long to look for the whole bricks he should have to build the mould in a reliable manner as if the pile was filled with the broken pieces; and it takes as long to lay a half of a brick as it does a whole one.

A loam moulder should be as careful of keeping his bricks whole and in good shape as a green sand moulder should be to keep good flasks in order to make reliable moulds and good castings.

A loam moulder that takes pride in having his bricks keep as whole as possible, will train his helpers so that they can have ready for him what sizes and pieces of brick he may require as he goes along, and any one breaking a whole brick not called for should receive a reprimand from him. When building some loam moulds it is as essential to have halves and pieces of bricks as it is to have whole ones; for as to halves, we are sure of having plenty of them to work with, and the manner in which they should be used in building up loam moulds is represented by the cut *B*, which is an 8″ wall, and a section of the outside part of a cylinder casting; the inside bricks that the loam is rubbed on to make its surface smooth, is built with the halves and pieces of bricks, and the outside is entirely built of whole bricks, care being taken that all joints are broken evenly; on the top of every 6 or 4 courses of the two 4-inch walls there is built one row of headers of whole bricks, as shown at Nos. 1, 2, and 3.

In building a mould after this plan, it can be carried up in most cases as high as 5 or 7 feet without the aid of any iron plates or rings to hold the brick-work together, which it would be necessary to have if nothing but pieces and a few whole bricks had been used to build the mould with. If a core was being built up with an 8-inch wall, the order of the half and whole bricks should then be reversed from that shown in the cut, so as to bring the halves and pieces to the part of the mould that forms the shape of casting.

There are two reasons for having the pieces of bricks next the surface of the mould which encounters the hot melted iron. The first is, that halves and pieces of bricks, when built in a circular form, will result in an evener thickness of loam all around the mould than when whole bricks are used; and the smaller the diameter of a mould, the more necessary it is to build this part with halves and pieces. The second is, that halves and pieces allow more joints than whole bricks, and thus afford more openings for the gases and air confined in the loam to escape through.

When bricks are built for thick or thin casting, there should be a difference made regarding the openness of the joints, the thinner the castings the more open they should be, so as to allow the gases and air in the loam to escape as quickly as possible, in order to prevent that cold shut and rough skin that some thin castings have, which is more fully explained in the article entitled MIXTURES OF LOAM.

The joints of bricks are in fact best when made open in almost all classes of work, whenever it can be done without danger of having cores bursting or swelling on casting.

There are ways that joints can be built open, and still reliable; one is, after a layer or course of bricks is laid, to pack well all the joints with the mud that you use for laying the bricks with.

The way to tie or build a square or corners of a mould is shown by the cut *D*, and outside corners like these require more caution when building than almost any other part of moulds, since the least weight or a knock will cause them to tumble down, if not well tied, when they are built very high.

The cut *H* shows the reason why some castings have been lost by the core giving way at the bottom of the mould, or at a point where the pressure of the melted iron found a weakness in the brick wall. The bricks were laid in this case without any regard to breaking joints either on inside or outside walls, and whole bricks were laid at random, sometimes going half-way around the outside, and in the next course the whole ones would be used to complete or make a part of the inside circle; while combined with this unskillful and unsystematical mode of brick-laying, the joints or openings between the bricks were not packed or filled up with mud as they should have been. If the moulder had only used the whole bricks he had laid at random, for building up the inside courses, and kept all the halves and pieces for the outside courses, and packed between all the joints solidly, he would not have had his core burst in when the pressure of iron came on it.

The cut *A* shows the way that a casting appeared made with an 8″-wall, at the bottom portion *X*, and from which it commenced to sag at the top, a 4″-wall was only built; the laying of the bricks in the 4″-wall was not done in a reliable manner; the joints must have been left unpacked and some distance apart. In looking at this casting the wonder to us is that the core did not give away at some point when being cast, so as to let all the iron run out of the mould, and it would most likely have done so had the core been built with halves and pieces; but it so happened that it was built with a new batch of whole bricks, and the joints being broken well when building saved the core from bursting in. Why the casting did not swell at the bottom and top, was because of the 8″-wall at the bottom, and at the top the pressure was not sufficient to squeeze the core in. *W* shows the way the casting would have looked had the core been correctly built by having all the joints well packed; and as this casting shown was an actual occurrence, I could not think of anything better to show the results of improper brick building, and to prove that in this, as in everything a moulder has to do in order to make a good casting, there is a right and a wrong way of working.

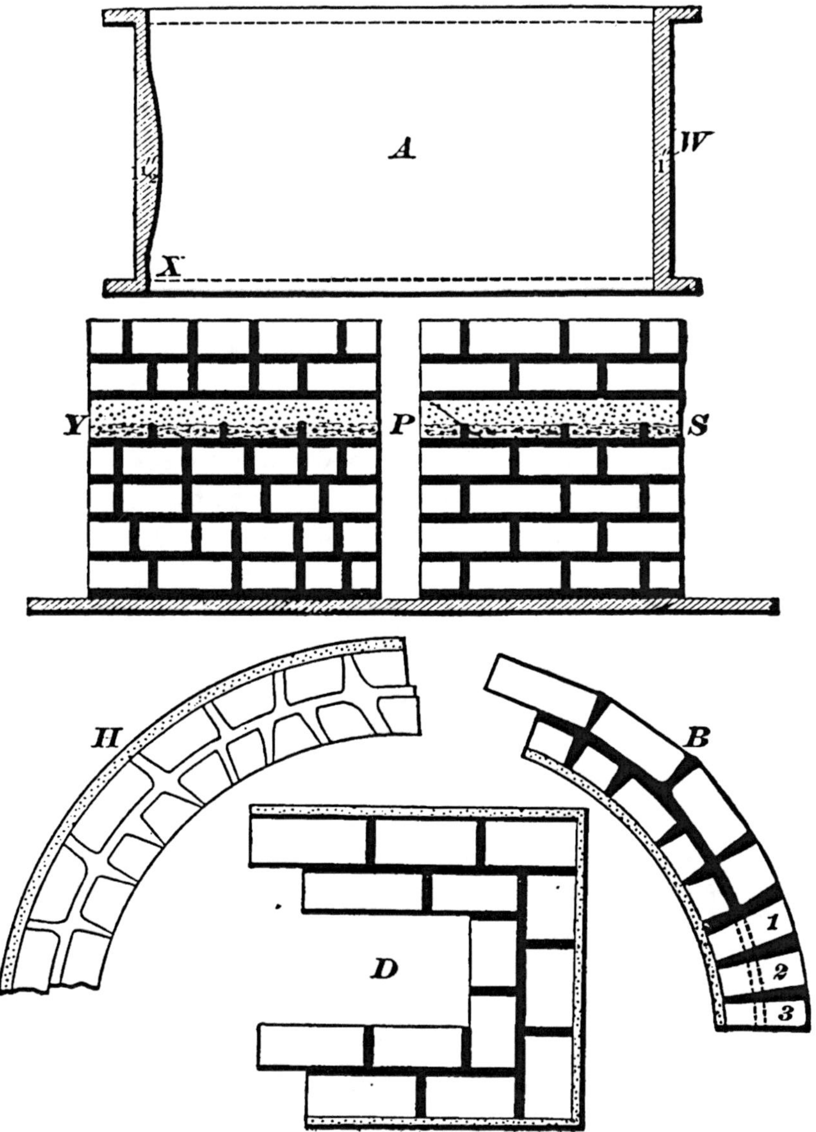

VENTING LOAM AND DRY SAND MOULDS.

LOAM or dry sand moulds require that some parts be vented more than others. There are often castings made in moulds that are never vented in any shape or manner. This is no proof that everything can be cast in loam or dry sand moulds without venting. There is not nearly the percentage of venting required for dried moulds as for green. In green sand moulds there is steam to contend with, which is not found in thoroughly dried moulds. A plain dry sand mould having 8″ of sand between the pattern and flask would be cast with less danger of its scabbing, if not vented, than if there was only 2″ of sand. A good body of sand in a flask allows the surface gases (and steam, if any) to have a chance to confine themselves in the interior body of the sand; and when there are no holes or openings in the flask, it may remain there, for there is room enough to hold the gases, and their pressure will be insufficient to force them through the face of the mould. But when there is not sufficient of sand to hold the pressure of the gas, it will obtain relief by coming to the face of the mould, and pass up through the liquid iron, causing scabs on a casting, *so that as the body of sand increases the pressure of the gases decreases;* that is, when there is no allowance made for gases to escape by venting the moulds or by the flasks having holes in them; also, conversely, where there is the least body of sand the more venting will be required. As a general thing, plain castings can be made in dry sand without being vented, when the flasks are in sections or jointed together, as joints or sections leave openings through which some of the gases can escape.

It is only in that portion of the surface of a mould which becomes heated to a high temperature that gases are formed or created. The surface of a loam mould has in many cases a better chance to be relieved of its gases than an unvented dry sand mould, on account of the openings or space existing between the joints and courses of bricks. A rod or stick rammed up on the outside of a loam mould will carry off the gases from a larger surface than one would rammed up in a dry sand mould, since the body of the loam mould is more porous.

When building up brick-work for loam moulds, the joints can be left open more or less, so as to allow the surface gases to escape backwards freely. Oftentimes it is necessary to have brick-work built very solid, so that the pressure of the iron when poured will not burst in or out the brick-work or cause the casting to swell. There are different means used by the various shops to accomplish this, and still provide a way whereby the gases can escape. Some will use straw between the layers or courses of bricks, and others will build their brick-work very open, and fill up between all joints with cinders, rammed in solidly, using a file or thin piece of iron to ram with. Again, some keep a sufficient thickness of mud between the courses or layers of bricks, and then vent between the courses of bricks with as large a vent wire as will possibly go between them; while again, there are shops that will apparently build up their brick-work without the use of vent wires, straw, or cinders. Such shops generally have a very open mixture of mud and loam to work with; it is very seldom that the joints of brick-work require to be so compactly built as not to have some porousness among them. A good plan to adopt where both good venting and solid building is required, is to compactly fill up half the thickness of the joints with mud, and the upper opening with cinders. Round loam cores, such as are used for forming the inside of cylinders, etc., are a class of building that must be the most solid; and the reason that such cores often stand closer building with less venting than the outside portion of the mould is, the core brick-work is not rammed up with moulding sand like the outside part of the mould, and the brick-work being exposed, the gases can escape more freely; the more open the joints, although compactly filled up with mud, the better chance for the gas to escape, since the mud used between the joints can be made of an open mixture, so as to be far more porous than the bricks used. Hard bricks should never be used to build up the face of a mould, on account of not allowing as free a passage of moisture and gases through them as a good soft brick. A good loam moulder understands what part of his mould requires to be vented, and also what parts will receive no damage if not vented. It is the same when venting loam moulds as with green sand moulds; flat, horizontal surfaces, corners, projections, and flanges require to be vented, while plain vertical sides of a loam mould can be built up without any provision made in many cases for venting. To take the gas or vent off from pockets, projections, corners, etc., in loam moulds, the use of straw, rods, or pieces of ropes or strings built in with the loam and brick-work, are generally used. When ramming up a loam mould to be cast, rods or sticks are laid about every two feet apart, against the brick-work, and when there is a flange, or any part of the mould that requires to have the vent taken off from it, cinders are connected from them to the upright vent rods. Should the vents be any special core vents, or ones that would be apt to make trouble if they get smothered, it is a good plan to cover the cinders with some paper, so as to keep loose sand from mixing in with them. For the plain parts of a mould, if cinders or straw are put around the mould at the height of every foot, the gases or vents will generally find their way out. The cinders or straw should be connected with the upright rods or vent sticks. Although venting loam moulds is looked upon as a small part of the work, it must be done with intelligence and understanding.

MOULDING ROLLS AND MAKING ROLL FLASKS.

Rolls for rolling mills are often contracted for by foundries that never have had any experience in their manufacture, and their success will depend upon the workman's ability and the foreman or melter's knowledge of making the right mixture of iron. The castings may be very rough looking and nothing said about it, but let the grade of iron be wrong, and the growling will begin. There are hardly two firms that will be satisfied with the same grade of iron.

I shall never forget the situation in which I was once placed trying to please two masters. One of these was the superintendent of the mills, and the other was the roll-turner, who had the turning of the rolls by contract. The turner, to a great extent, had his say as to what foundry should make the castings, and to get the work it was necessary to have the castings soft enough to suit him. Occasionally the superintendent would give us a call, and want to know how it was that the rolls could not be made harder so as to wear longer, and give us to understand that when So-and-So made them they lasted a great deal longer. Our only answer would be that we supposed they were all right, as the roll-turner had not said anything against them ; then for a while the rolls would be made harder until the turner would commence to growl again.

As a general thing the purchasers of roll castings like to

have them as hard as they can be without having the edges or corners chilled.

About the first thing required in starting to make roll castings is to have a flask to mould them in. The style or shape of the flask will depend upon the way that they are to be moulded. The old style of moulding such castings, and one that a great many shops yet follow, is to have a full pattern for every shaped roll wanted. To make such patterns costs much time and labor, and a large warehouse in which to store them away.

I am an advocate of the more modern plan of making rolls by sweeping them up. The cut shown is a flask intended for that purpose. I have seen many different kinds of flasks for such jobs, but the one shown has, I think, many good features. There is one point especially that I claim should be provided for in such flasks in order to make good, smooth castings, free from scabs, and that is to have plenty of vent holes cast in the flask, whereby any gas or steam is allowed to escape. As a rule, there is no allowance made for gas or steam to escape, and if you should ask the moulder that designed the flask why it was so made, he would tell you that it was for a dry sand mould, and therefore it did not require any vents.

In the end of the flask shown will be seen 13 one-inch vent holes, and the same in the cross bar. When making the roll long, $\frac{3}{8}''$ or $\frac{1}{2}''$ rods are inserted so as to run the entire length of the flask. When the mould is ready to be blackened the rods should be taken out. With this system of venting, when everything is done as it should be, you can rely on having a good casting free from scabs. Such vents also greatly aid in the drying of the mould. One trouble often experienced in making such castings is having to provide for making large and small castings in the same flask.

When the casting is small there is generally a heavy body of sand that the bars do not assist in holding in, when the cope is rolled over, and this hanging sand is liable to drop out. Again, the bars having been first made right for the medium-sized castings, the first thing we know there comes along a casting that is too large in diameter to be admitted between the bars. Then the bars must be chipped out.

A good way to get over this difficulty is to make the bars so that as large a casting as should be made in the flask can be admitted between them, and then, when a sweep or pattern comes along that is so small in diameter as to endanger the dropping of the cope, false bars, as shown, can have pins *P*, *P*, inserted and wedged in the cross-bar holes *H*, *H*. This plan will, I think, be seen to be a better one than driving in a lot of rods or gaggers to hold the hanging sand, as is usually done.

Another point that may be noticed is the plan here shown of using loose plates instead of a large, clumsy back plate, and a lot of bolts or clamps. Of course a back plate would help to make a flask stiffer, and in the case of making extra heavy rolls, I would recommend its use, but for rolls weighing from three up to eight tons, the loose plates are safe.

Sometimes roll flasks are made with the bars, sides, and ends all in one piece. This plan I do not approve of, as it not only costs more to make the pattern ; but when a flask is thus made, there is more or less danger of its cracking, and when there is a serious break, the whole half flask has, generally, to be broken up and a new one made. When a roll flask is made in sections and bolted together, there is only one side, one end, and one bar pattern required, and a flask thus constructed can be made longer or wider at any time if so desired. Should any parts crack, they can be readily replaced. The handles generally used for this class

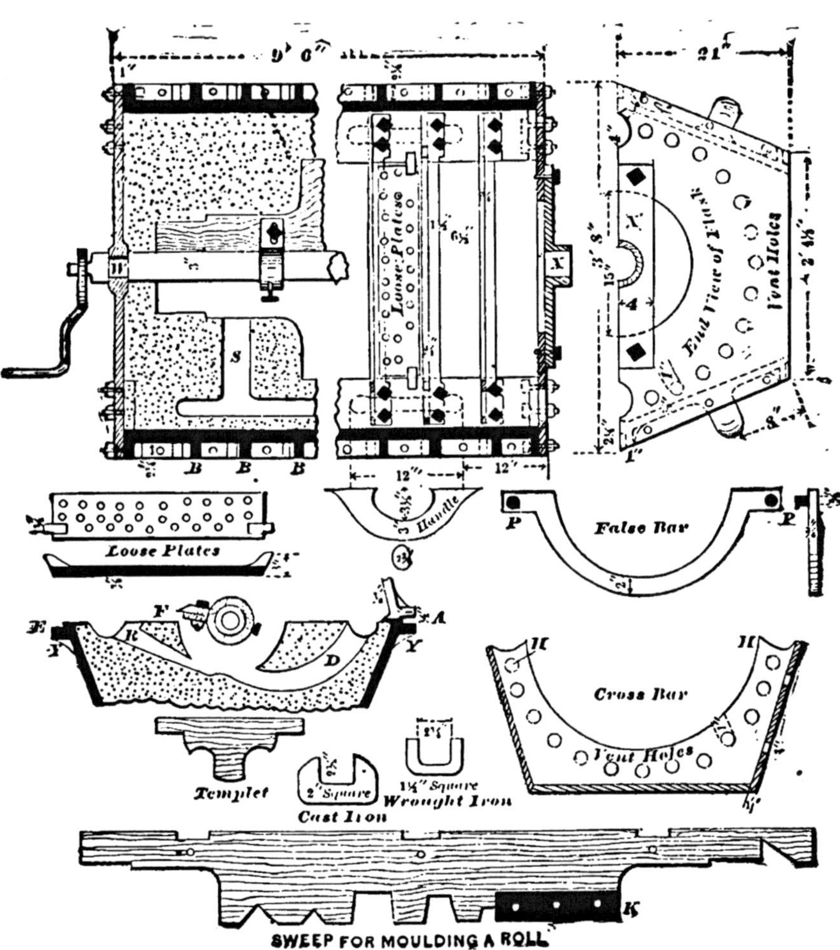
Loose Plates
Handle
False Bar
Cross Bar
Vent Holes
End View of Flask
Templet
Cast Iron
Wrought Iron
SWEEP FOR MOULDING A ROLL

of flasks are cast iron, and, in order to be strong enough, they often look very clumsy or out of proportion. Trunnions are sometimes cast on the ends of the flask to roll them over by. In the cut there are only two of the four handles shown.

About the first part of a flask that gets broken through usage is the flanges, and often castings have been lost from the flange breaking when the mould was being poured. At *E* and *B, B, B,* is shown a reliable plan for constructing flanges so as to stand the repeated strains they are subjected to. *Y, Y* are brackets that give strength to the flanges, while *B, B, B,* being level with the rest of the planed joint when the two parts of the flask come together, will prevent any straining or springing of the flanges when the bolts or clamps are used to hold the two parts together. The space of one half-inch, as shown at *A,* is to leave room to pack sand or loam between the joints to prevent any running out. Such flasks, after being used a few times, will warp more or less, and, although the joints of the flask were planed so as to have a good bearing, the reheating of them will soon make it necessary to pack them.

To fasten the two parts together one shop will use bolts, while another will use clamps. Either way will answer the purpose. A bolted flask is safer than a clamped one, the only objection to the bolts being the trouble of unscrewing them, and keeping the sand and dirt from destroying the threads. The objection to clamps is the jar given to the mould in hammering and wedging to fasten them.

Among the cuts will be seen measurements for the making of cast or wrought clamps, such as are used for ordinary roll flasks. Wrought iron clamps are safer than those made of cast iron. A flask should have more bolts or clamps on the end that is cast down than on the upper end, because there is more strain on the lower end. In such a flask as shown, the bolts or clamps should average 6″ apart at the lower end, and 8″ at the upper end.

When possible, it is best to have four or six bolts put in to assist the clamps. I have seen flasks, when the pins were taken out and clamps put on in their place, get a jar when being hoisted up on end so as to loosen the clamp and cause the two parts to shift. Clamps should never be all wedged in the same direction. Each alternate one should be wedged the reverse way.

When fitting together a flask for sweeping up rolls, the joints, if not planed, should be chipped so as to fit closely together; then the three or four pin-holes should be drilled, and the end bearing, *W*, bored out. The spindle holder, *X*, should be accurately fitted with set screws before the flask is taken apart.

I once worked in a shop where the moulders did not use any pins to close the flask by. They would use the bearing, *W*, for a guide to close the lower end by, and for the upper end pass the arm through the riser head, and feel the joints of the mould. If not right, a man on each side of the flask, having a sharp flat bar could easily move the flask as wanted. Iron wedges are placed between the iron joints to keep the joints of the mould apart and save crushing. When the inside joints of the mould are even with each other, the wedges are taken out and the cope let down to place.

The cut in which the handle is shown is for illustrating the process of sweeping up a roll, *F* showing an end view—the spindle shown is a tube with solid ends forged or cast on to work on the end bearings. The idea for thus making it was to have it light to handle.

Sometimes instead of having a journal turned in the spindle, so as to keep it from working endways, as in the one shown, there are two collars fastened with set screws to the spindle, so as to have one on the inside and one on the outside of the lower end of the spindle bearing.

In sweeping up rolls the sand is not all knocked out of the flask when a casting is made, as is done when a full pattern is used. After a casting is taken out of the flask only the loose and burnt sand is taken out, or enough to allow of from 2″ to 4″ of tempered loam or dry sand being rammed in to sweep up a fresh mould. In ramming the sand some moulders use only their hands, relying on the extra dampness of the sand to make the mould solid enough.

This is a plan that I do not approve of. I know from experience that a better casting can be made by working the dry sand but very little damper than green sand is generally made. To have the mould solid, use a rammer instead of the bare hands. In order to have this, three or four inches of fresh sand adhere to the old sand, there is a coat of thin clay wash sprinkled over the surface of the old sand, and then a coating of mud rubbed over that. On top of this the tempered sand to form the mould with is shoveled in.

Sometimes instead of having sand between the bars the space is packed in a reliable and solid manner with fire-brick, and then every time a casting is taken out all the sand is removed, and the mud rubbed on the bare bricks, for starting a new mould. This plan is a good one, where a shop has three or four different sizes of flasks to accommodate different diameters of castings; but for a shop that has only one flask I would not advise its adoption, as there would be sure to come along some sweep that would require nearly all the bricks to be cut out to admit it, which would be sure to loosen the under bricks.

The sweep shown for moulding a roll has in it the square and the half-diamond-shaped grooves, such as are generally used in rolls. The diamond grooves are easily swept up, but the square ones are more difficult, and often require to be well rodded in order to stand. When sweeping up the rolls the grooves and surface of the mould are swept up as full as can be with the dry sand mixture, and the surface of the mould is made smooth by using two coats of loam, the last one being about as thick as buttermilk. The beveled edge of the board only is used for the finishing coat, which should be accomplished in once going around, and as quickly as possible ; that is, the board or sweep should be turned slow and steadily, but putting on the loam with a brush must be done so as to lose no time. The striking off of the joints is done by having a straight-edge work lengthways of the flask.

The joint sweep, also all the roll sweeps are better for having the working edge of sheet iron, as when they are all wood they soon get worn out. The sheet-iron plates can be fastened on the wooden sweeps with screws, as shown at *K*.

The templet shown is for a guide to set cores by to form the wobblers on the roll. At *S*, *D*, and *R* is shown the plan generally adopted for the gating of such castings. *S* shows the part of the roll to which the gate is attached, and *D* and *R* show two different forms of gates used to cause the iron to whirl around as the mould is being filled up, so as to bring the dirt to the center and keep it from being lodged under the grooves.

Rolls are always cast vertically, and the hotter and faster the iron can be poured in, the cleaner will the casting be when turned up. The roll flask, and also the iron casing rigging shown on p. 217, was made by *Mr. William Fitzsimons*, of Cleveland, Ohio, a skillful moulder, and one having large experience.

THE SURFACE OF A LOAM MOULD.

To have a good loam mould a good surface is essential, and a good surface depends upon many conditions. First, the mixture of the loam must be correctly made; secondly, the loam must be put on in a reliable manner; and, thirdly, it must be finished up properly. Loam in some foundries is mixed up in mills which are made by taking a large flat bottom iron pan, from 4′ to 8′ in diameter, and placing a similar pan over it; and in these pans are two heavy grindstones. The pans are so made that they revolve the stones as they revolve, and the loam mixtures are shoveled into the pans with its water or clay-wash to wet it with; the heavy stones rolling over the different parts serve to unite and mix them. Another plan sometimes adopted is to beat the mixtures of sand, when wet, with a rod of iron, the loam being on a wooden or iron bench. Of these two plans the mill is by far the best, in fact some shops would never think of using a loam unless so mixed. The different mixtures of loam used are many, most every shop having a different mixture. In some places a natural loam can be obtained—but this is rare; most shops have to make their loam of different proportions of sharp and loam sands. There are certain conditions or qualities that should exist in all loam mixtures alike. If a loam mixture which produces good castings in any other foreign foundry were brought to your shop to be used, and was handled in the same manner, it would be your own fault if you could not turn out as good castings as the foundry from which the loam came. A good practical loam moulder can tell by feeling of loam when mixed if it will work well or not. A sharp sand is used to regulate the loam sand. The more clayey or loamy the loam sand is, the more sharp sand must be mixed in with it, until the practical moulder is satisfied with its consistency. A simple way to try a new loam mixture is to take a lump of it, after it is well dried, and immerse it in a ladle of iron. If the iron boils after the first bubble, the mixture is generally too close or clayey. Loam should be of a porous, but firm nature ; if it is too porous, on the other hand, the mixture will crumble to pieces by a gentle squeeze of the hand. Loam should be mixed weaker for castings below one inch in thickness, than for those of a greater thickness. A loam that is strong enough for a casting four inches in thickness, is strong enough for any heavy body or any thickness of iron. What is meant by loam being stronger, is, it is more close and clayey. A heavy thickness of iron will scab, just as a far lighter thickness will with the same mixture of loam if it is too close or clayey ; the thickness for loam put on bricks, to form the surface of a mould, should be regulated by its form. A thin thickness of loam is more liable to cause a mould to be scabbed than a heavy thickness. Loam should not be put on any less than $\frac{3}{8}''$ up to $\frac{3}{4}''$ for plain surfaces of moulds ; but for pockets, corners, and flanges, etc., loam should be no less than 1″ in thickness. Burnt or hard bricks should not be used to form the surface brick-work of a mould, or for corners and pockets, or any portion of a mould's surface that is liable to scab. Moulders very often use what is called loam bricks, or cakes, instead of using the common ordinary bricks. The loam bricks are used on the principle of *the thicker the body of loam, the better chance for the gas to escape,* and thus cause the iron to lay more kindly against a mould's surface, also to allow for contraction of the casting when cooling.

To make loam bricks, use a loam as coarse as that used

for rubbing on the surface of the mould, and have a wooden frame made for whatever shaped bricks or cakes are wanted. Then set it out on oiled iron plate ; fill up the frame with the loam, having it mixed as stiff as will work easily. After the plate is full, the soft bricks are then set in the oven to be dried. Sometimes whole cores and large portions of a mould are formed with loam bricks.

The first coat of loam that is rubbed on the bricks should be the openest, and to finish up or form the face of a mould use finer. Before the fine finishing coat is put on, the face of the mould should be swept up as full as possible with the open loam, as the least amount of fine loam that can be used, the less danger there will be of a mould scabbing. With most mixtures of loam it is best to have the finishing coat put on as soon as the rough coat becomes stiff enough to hold the finishing loam, and have one or two revolutions of the sweep to make a finished face. When the first or rough coat is allowed to become hard or air dried before the finishing coat is put on, it will not unite or cement as well as if it is put on having the rough coat as above described.

There are two ways practiced in making finishing loam mixtures ; one is to use the mixture as for the rough loam, and have it put through a fine No. 8 sieve ; the other is to use some foreign mixture (for receipts and mixtures of loam, see notes and receipts in the back part of the book). While the finishing coat must be fine, its mixture should not be close or clayey, and it is better, if possible, to obtain a finishing loam mixture, whereby the face of the mould can be made smooth enough to receive the blacking without the use of tools, as the less sleeking done, the less liable a mould is to be scabbed. This only refers to swept portions of a mould ; for other parts that patterns are used for, there is more or less sleeking done with tools, that cannot be avoided.

SWEEPS AND SPINDLES.

SWEEPING green and dry sand or loam moulds is a branch of the moulder's trade that in general calls for higher mechanical qualifications than making castings from a full pattern. Sweeping or bedding in is not extensively done in other than jobbing or machine-shop foundries, hence only a comparatively few moulders are acquainted with the processes; but since the practice is becoming more common from year to year moulders will be required in the future to give more attention to this part of the trade.

In the cut is shown a rigging for sweeping under the bottom of loam cores. The sweep, seen at the left, is bolted to two iron arms *X*, *X*, which are held up by two collars fastened to the spindle with set screws. The sweep revolves around, and the spindle remains stationary. The tapering end of the spindle is set into the casting *H*, the outside diameter of which can be from 4″ up to 8″. This casting should be turned up on the outside, true with the chilled inside spindle hole, so as to have a true surface for the lower arm and collar to be placed and worked at any point up or down on it. This casting is bolted to a plate from four to six feet diameter, and the plate is laid level on a solid floor. The loam plate is then set on top of the spindle-holder *H*. Bolted to this loam plate is a casting having an upper and a lower flange. In the upper flange there are four staples cast, two of which are shown at *E*, *E*. The inside of this casting is bored out the size of the spindle, and when bolted to the loam plate, as shown, and the spindle passed through it, there is no danger of the loam plate being overbalanced. Another plan, which would in many cases be better than to bore out this double flanged casting the size of the spindle as shown, would be to cast a hole in it about one inch larger than the size of the spindle, and by having three set screws near the bottom, the top could be fastened with pry wedges. By this plan there would be a better chance to regulate and level loam plates, and also it admits of the spindle being put in and removed more easily.

Before setting this loam plate on the standard *H,* it is daubed up with loam even with the face of the prickers, and then dried in the oven, so as to have a dry body to absorb the moisture of the loam used to finish it up with when the plates or bottom is swept up as shown.

When this spindle-holder or standard is used it is generally for a large core that has little or no bearing on the bottom of the mould, but has to be supported from the top as in the cut. The top loam and covering plate is not set on and bolted to the lower plate until the core and a level joint is made and finished with the sweep. This top plate having been previously swept level and dried, requires no sweeping to make it have a true face after it is placed.

After this plate is bolted with four bolts, one only of which is seen, there is a row or two of bricks, *P,* built around on top, and a sweep forms a straight face the same diameter as the one swept on the outside or cheek *W,* so that when the core is lowered down into the mould, a short straight-edge placed against the parallel faces *P* and *W,* will center the core in the mould.

The staples *E, E,* and *A* are for hitching the chains to hoist the core by. The top staples, of which there are four, are the best to hoist by, but should the mould or core be less in height, the lower staples can be used. When the core is

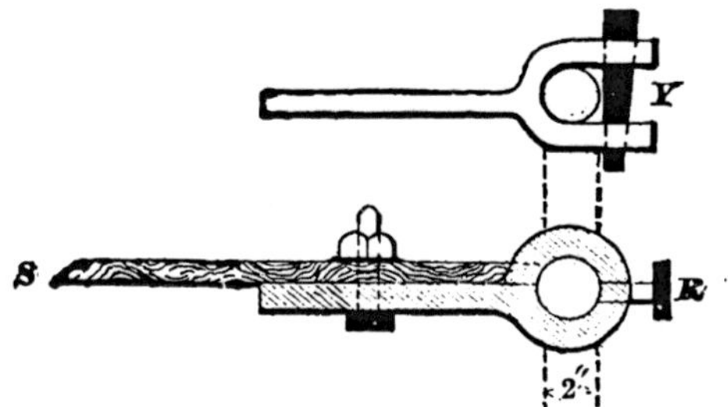
Y
S
R
2"

finished and hoisted up from the standard *H*, the hole that it leaves in the bottom is filled up and made level with bricks and loam, a piece of plate iron having been first wedged in up against the spindle hole,so that the pressure of the melted iron cannot burst through the bottom when the mould is poured.

When setting this core on the carriage, the bottom is lowered down on a flat plate having on it a bed of dry sand for the bottom to rest on.

For other classes of work, where all the bottom is not wanted, a false ring or spider having a hub 6″ long with a hole bored equal to the outside diameter of *H*, with an inside flange which is for resting on the top of *H*, could often be used to a good advantage as a supporter for loam rings, etc.

This rigging, for a jobbing shop that does much loam work, can often be used to a good advantage for casting short stroke cylinders that have one of the ends cast in, as it is now often done; by the plan as shown the bottom could be cast down, if desired.*

To form a riser head on a cylinder when this plan is used, the top flange is bricked over and the straight part of the cylinder carried up as high as wanted. When making a cylinder this way the outside should be checked off, and when the cores are set in and fastened, the center core is lowered in. A man with a lamp underneath can guide and see that the center core does not touch any of the port cores. Then the whole mould is set on the bottom by hoisting it by the four cheek handles, one of which is shown at *T*. This bottom joint should be made beveling, instead of straight, as shown.

The gate shown is for filling the bottom over so that when the iron drops down from the top it will fall into iron and not cut the bottom of the mould.

The cuts *Y* and *R* show two styles of arms. *R* is a style that can be worked tight or loose. When tightened on the spindle by the set screw shown, the spindle must revolve;

* A rigging somewhat superior to the above for "under surface sweeping" is illustrated on p. 66, Vol. II.

but when there is a collar screwed to the spindle, as shown at *D, D,* the arms are loose and can revolve without turning the spindle.

The latter is the best plan when sweeping with a spindle that works in a tapering hole long enough not to require steadying at or near the top; but when a spindle is held at the top, and the bottom works in a small socket, as shown in the octagonal loam mould cut, there is very little friction, so that the spindle can be turned when sweeping very easily.

The arm *Y* is a very handy one to use on a spindle that is made to revolve in sweeping. This arm can be made for one or two keys, but it is best to have two in one to be used for holding heavy sweeps. The advantage of such arms is that they may be taken off and on without disturbing a spindle held at the top.*

The placing of arms or brackets for holding the top of long spindles steady, is an important detail that is very seldom properly attended to. There are very few buildings but that whenever a crane is turned around will move more or less, and in some shops the loam moulder when sweeping up a long mould has often to sit down and wait until a crane can be turned back the same as when the first coat of loam was swept on. Arms or brackets should not be fastened to unstable buildings, but should be secured to upright timbers sunk deep in the ground, and independent of the building altogether. The board or sweep bolted to the arm *R,* is to show how arms should be made. There are shops that have arms made so that a sweep when bolted to them will not have the working edge, *S,* on a true line with the center of the spindle. This causes trouble in setting the sweeps and getting the right diameter for a casting.

In making spindles they should be made even inches diameter, otherwise they are apt to cause mistakes in making

* Another good point for "spindle arms" is given on p. 68, Vol. II.

and setting sweeps. $1\frac{3}{8}''$ $1\frac{7}{8}''$ or $2\frac{1}{8}''$ makes trouble for the moulder as well as the pattern-maker, as it is apt to confuse. Two inches diameter makes a handy spindle for ordinary sweeping, and for fine work they should be made of steel, turned up true. The larger sizes of spindle are often made of wrought-iron tubes, or of hollow cast iron.

The spindle-holder for sweeping green sand moulds, that has been shown so many times, but never explained, is a flat plate about 24" diameter, and the tapering hole for the spindle to fit into is about 10 inches long.* When casting this, the tapering end turned on the spindle can be used for a chill, being set in the open mould and the iron poured around it. While hot the spindle is knocked out, and when put in again, to use for sweeping, you can rely on having a steady spindle. The collars should be used on this spindle, so that the arm and sweep can revolve without having the spindle turn. This makes a very handy rigging for sweeping green sand moulds, as the spindle seat or holder is light, and can be quickly set in any part of a foundry floor.

The tapering end of this spindle should always be well oiled before it is set in, as otherwise the damp sand and steam are liable to rust it.

As the sweeping of green sand mould is generally done to save pattern-making, the proprietor, as well as the moulder, has the advantage over others when he can make a casting with sweeps that others could not make without having a full pattern to work with; and in cities or places where competition is active, a good knowledge of sweeping, in all its branches, will be of value to the proprietor and moulder alike.

---

* A "spindle-holder" of the above style is illustrated at B, p. 69 this book.

MOULDING GEAR WHEELS IN DRY SAND OR WITH CORES.

To have smooth, even teeth is a very important feature in the manufacture of gear wheels, and the most reliable way to make good teeth on large wheels is to have them moulded in dry sand, or with cores. Often large wheels made in green sand will have teeth larger than the pattern, as the sand will yield more or less, depending on the way it is rammed. Although there can be nice-looking gear castings made in green sand, the same pattern moulded in dry sand will make a casting that will run easier and quieter, and wear longer. A variety of spur wheels could have the arms and hub moulded in green sand, and the teeth in cores, or dry sand, by having an iron ring or flask to carry them, so that the teeth could be hoisted and placed on the oven carriage to be dried. If there is not a full pattern to mould the wheel by, a segment could be used the same as is shown for sweeping up gear wheels in green sand.

The sand should be closer and tougher for ramming up a spur wheel than for ramming or bedding in a bevel wheel; and should the same close sand be used for the bevel as for the spur wheel, the teeth will be very liable to scab. Dry sand in this respect is the same as green sand, as the sides, or any part of a mould that the iron rises up against, will stand harder ramming, and will require less venting than the bottom, or any part that the iron lays over, or on the top of. It is a good thing that this law or principle is as it is; for if the sides had to be rammed as soft, and the sand left as open, as is required to keep the bottom or flat surfaces from scabbing, the side of some moulds would fall down, or cause a deal of extra rod-staying and other precautions to make them stand.

A very essential point in making gears in dry sand is to blacken the teeth, so that they will be smooth, and not show streaks, or lumps of blacking on them. In blacking teeth the blacking should not be thick, and, if a swab is used in order to quicken the operation, it should only be used for the first coat, and a camel's-hair brush substituted for the second and third coats. To make a good job of blacking requires neatness and care, and a moulder that takes a swab and pastes on the blacking, washes or knocks off the edges of the teeth by rubbing the swab against the mould when there is hardly any blacking on it, and then attempts to finish or patch the teeth by the use of tools, will make a very poor job. Teeth should be blacked with much care, and so smoothly that it is not necessary to touch a tool on them; for if a good job cannot be done with a brush, it cannot be remedied with tools.

There are very few shops that mix their dry sand or loam alike, for the reason that they have different grades of sand to deal with. A suitable mixture of open and close sand to form a loam, or dry sand, that will stand the fall and wash of the iron without scabbing, is generally arrived at by experience, although there are ways of telling whether new mixtures will work right, which is discussed in other parts of this book. Take any dry sand-facing mixture that works all right on ordinary castings, and mix it a little closer; put in one part of sea coal, coke, or blacking, with from twelve to twenty parts of sand, and it will help to make the sand peel, assist in making smooth teeth, and give them a good color.

This cut shows a good plan for making gear wheels with-

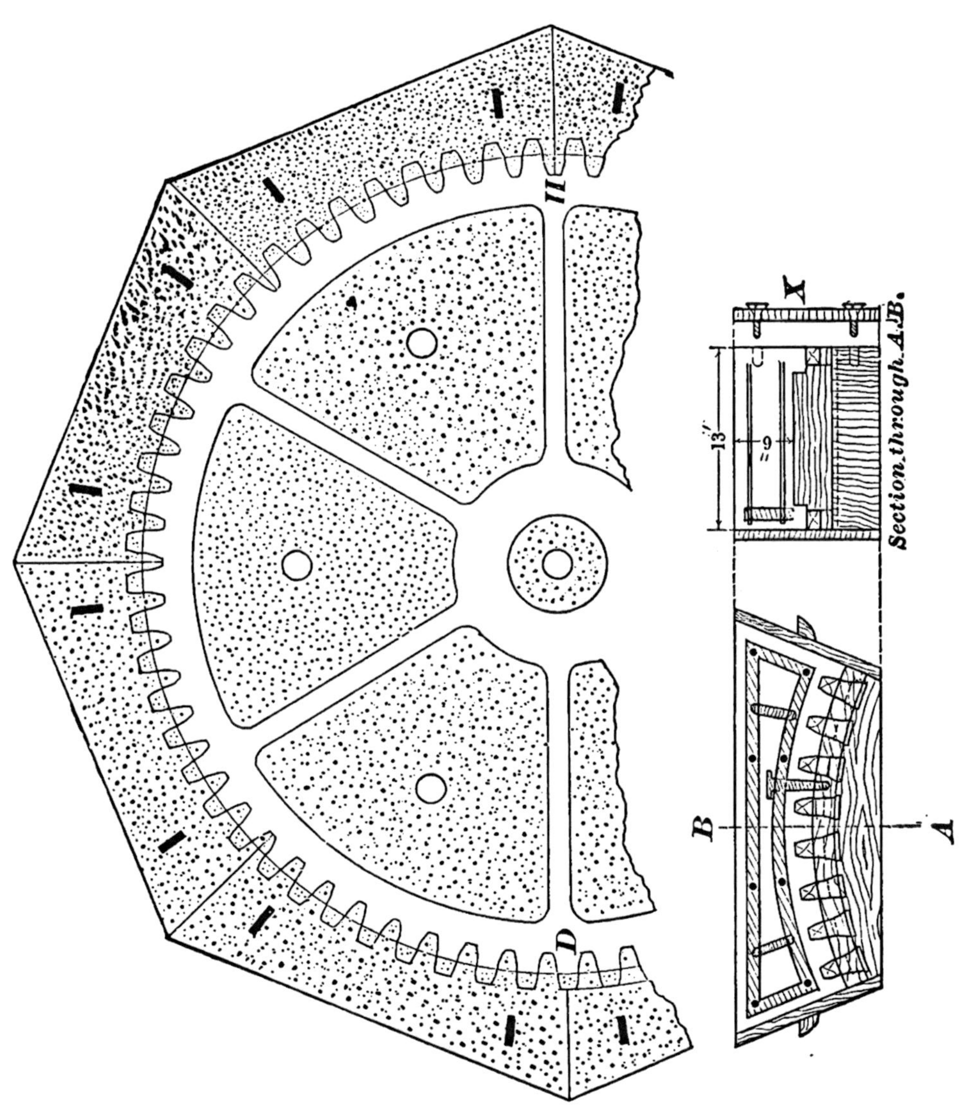
H
D
X
13″
9″
Section through A.B.
B
A

out a pattern, the teeth being formed from cores made in the core box, as shown. In the box is seen the cast-iron core frame, rods, lifting hooks, and spike nails for holding the teeth. The face of the box *X* is made to be taken off, to allow of drawing the remainder of the box without breaking the core. There is a shrouding on the top and bottom of the teeth, which is formed in the same box.

There is a tooth sometimes used in gearing that is the largest at the pitch line, so that it cannot be drawn out of the sand flatways like the one shown. For such teeth the core box has to be arranged so that the teeth can be drawn out endways before the core box is drawn off; and to form the top shrouding there will have to be separate flat cores made.

The arms of the wheel shown in the cut were made or formed of cores dried in the oven. They could have been made in green sand, but it was safer to make them of dry sand, as there was a deep strengthening rib all around the center of the arms. In moulding or forming this wheel, strike off a green sand level bed, and if the cope is a wooden one that needs gauging, make the level bed hard, ram the cope up on it, soften it up again and finish it ready for setting on the cores. Then, with the sweeping board attached to an upright spindle, strike a mark around on the bed the diameter of the inside of the teeth; raise up the sweep so as to clear the top of the cores, after which set around the teeth cores, using the sand mark for a guide. When these are all set in their places, screw a strip of wood on the sweep that will come down and clear the inside of the teeth. In sweeping around with this you can see whether the cores are exactly true or not. There are two things that will have to be watched closely in setting the teeth cores. The first is to have the cores set in a true circle, and the second is to have the teeth where the cores join together the same size as otherwheres, in doing which a pair of calipers are useful. It takes time and patience to get the cores to come exactly right, and they may require to be moved several times to get them so. The circle may be right and one of the joints wrong, to remedy which the circle must be made smaller or larger.

In making the core box there are two ways of splitting the tooth, as shown at *D* and *H*. *D* is the best way to make the cores, and *H* the easiest to get at the joints of the cores when set in the mould for the purpose of blacking and drying; which should be so nicely done as not to show in the casting in the least degree.

After the inside joints are daubed, close or daub up the outside joints with mud; ram up the space between the bank and the outside of the teeth cores with sand, so as to keep the cores from being forced out of their places when the iron is poured into the mould. The level bed for setting the cores on should be sunk below the level of the floor to the same depth as the face of the wheel, and after the cores are rammed around, go round with the sweep and see if the cores have been moved. If they are all right, take out the spindle and sweep, and set the center core in the print formed by the sweep. After this is done, set in the arm cores, dividing them with pieces of wood—two for the arms and two for the rim. When all is right, put on the cope, and secure the arm core vents before pouring.

---

* On p. 261, Vol. II., Mr. Harrison objects to making gear wheels with segmental cores. The author admits it is not the most reliable plan; but if such work is done mechanically and with care, good true castings can be made. At least the author has made large good true wheels by the above plan.

MAKING RETURN, ELBOW, BRANCH, AND T-PIPE CORE ARBORS.

The making of crooked pipe castings is generally expensive compared with the cost of making straight pipes, the principal feature that increases the cost being the cores. The cut on p. 200 represents the making of a core for hot blast return pipes, the plan being that of Homer Hamilton, of the firm of William Todd & Co., Youngstown, Ohio, to whom I am indebted for permission to illustrate it. The old plan of making these cores was to make them in halves and paste them together, the halves being made in a wooden core box, or swept up on plates. Sometimes these cores were made in two sections and butted together at *R*. When set in the mould in whole, the core irons would usually consist of wrought-iron rods, spliced so as to lap by each other three or four feet, and to get them out of the casting required time, and a good deal of pulling and twisting. Some shops, for such jobs, would cast some light bars, one being in each half, and break them to get them out of the casting.

The core bar shown is all cast iron. The straight lengths are made as at *P*, which shows the end view of the bar, and also the core box having a core in it. The round holes represented are the vents. The core bar at the rounded end of the cut is also cast iron, being made of short links held together by having the projection *X* set into an opening, or between two lugs, *E*, *E*, and the rivet passed through and riveted. These links have also a guide, *F*, *F*, cast on them, so as to let the links bend inward as far as required. If it were not for these, when the arbor was set as wanted, on attempting to hoist it up the two sides would close together. For holding the other ends of the straight arbors stiffly in place the arm *W* is used. The holes are for the vent rods to pass through when making the core. This arm is put on when the arbor is put together, and is not taken off until the pipe is cast. The straight arbor on the right is joined and fastened to the end link by two hooks and a center-pin. The hooks and pins are attached to the straight

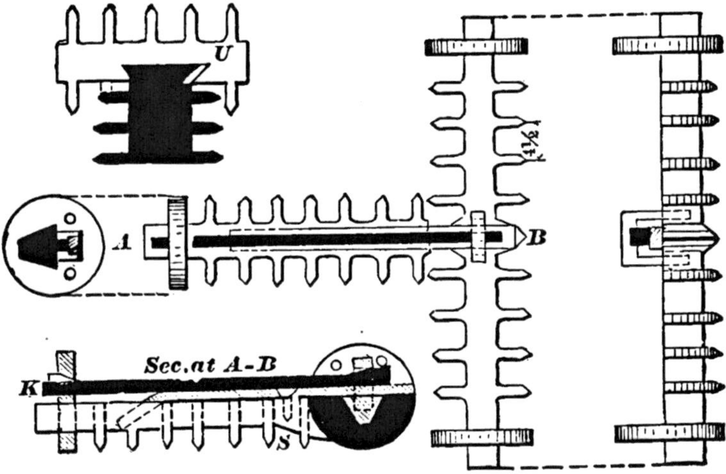

arbor, detached views of which are shown at *D* and *H*. At *D* is shown a side view of the center-pin and one of the hooks, and at *H* the end view and the sides of the arbor the hooks are bolted or riveted to. These hooks and pin fit in holes drilled into the end of the last link, and when taking the arbors out of the casting the set screws are loosened and the arm *W* taken off. The pipe casting is then hoisted up and pounded with sledge hammers until the sand is nearly all out, then the arbor on the right is given a turn in the

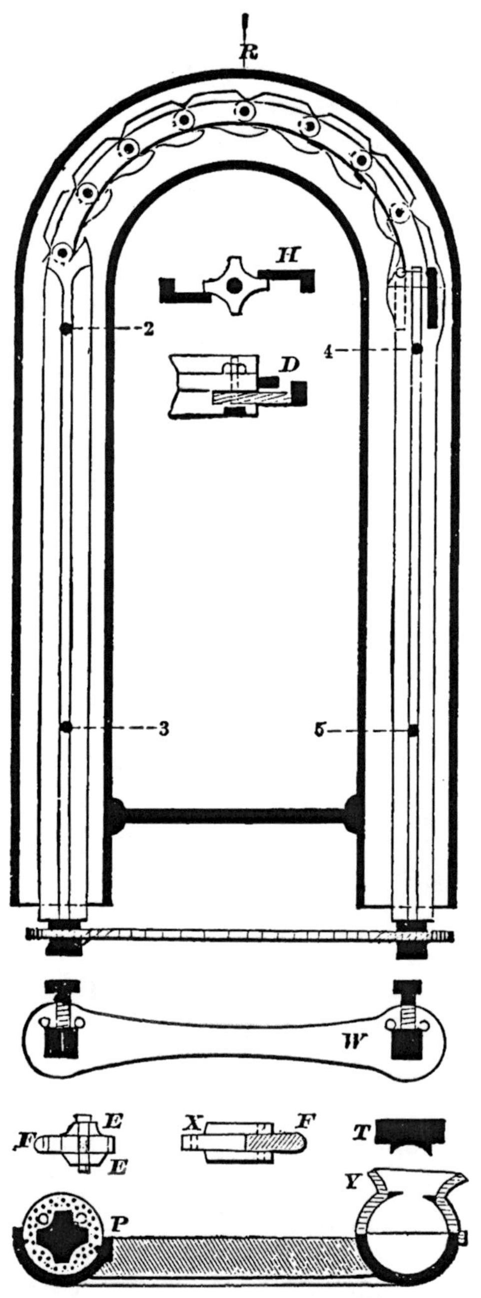

MAKING CORE ARBORS.

on, and a piece of a box about two feet long, an end view of which is seen at *Y*, is placed on top and the upper portion of the core is rammed up.

The small sweep *T* is used to shape the part left open, and through which the sand is shoveled in and rammed. This section, or upper box, is then drawn and replaced until the whole top part is rammed up. For the round end there is a short, circular box used.

The top half could be formed with a sweep, but ramming it in these boxes makes a more solid core.

The core, when dried, is set into the mould by four screws, screwed into the core arbors at 2, 3, 4, and 5.

The principle involved in the construction of this core arbor can be adapted to many other purposes besides making pipe.

The cut on p. 199, showing wings cast on a core arbor, represents two ways of making cast-iron core arbors for a large number of T, branch, or elbow pipes, using the same core arbors.

As a rule, such castings are made with dry sand cores, and if a cast-iron core rod is used, it has to be broken in pieces to get it out of the crooked casting.

With an arbor, as shown, the cores can be made of green sand, and the arbor taken out of the castings without breaking it.

The cut shows the arbor for moulding a T-pipe. The section through *A B* shows the branch part connected with the main or longest section of the arbor. *K* shows a wrought-iron square bar, one end of which is wedge-shaped. Underneath this wedge bar is a flat wrought-iron bar, one end of which is bent and cast into the arbor, as shown by the dotted lines. In this bar a countersunk hole is punched hot, as a hole drilled out would have a tendency to weaken it. Through this hole a bolt or rivet is placed and cast into the arbor, as shown at *S.* When taking the arbor out of the casting, the wedge bar is knocked in, when the arbor will drop out.

The cut showing a dovetail is a plan generally used to hold sections together. A hardwood wedge is driven into the opening *U*, and when the pipe is cast the heat will loosen the wedge and free the arbor. If either of these plans are not thought to be firm enough for very heavy cores, the dovetail and bar wedge could be combined so as to make a very stiff joint. Lengthways the dovetail should be quite tapering, so that when the pipe is rolled over, the wings on the arbor will allow it to drop sufficient to permit the dovetails to be pulled apart.

The cut and description of the complete arbor is not taken from any arbor, so far as known, in use; but the plan is one that I think would work well and be an improvement on the old dovetail arbor, which sometimes causes trouble by getting loose.

In order to make this class of work fast, it is as necessary to have good flasks fitted up specially for the work as it is to have good core arbors. The flasks are better if made entirely of iron, as they can be cast to suit the shape of the casting wanted, thereby saving much of the shoveling and ramming of sand.

If the flasks are made of wood, the ends should be iron having half-circle holes, so that the round flanges at the ends of the core arbor will fill them, making a bearing of iron on iron to hold the core up and down at each end. The flanges of the arbors should be no larger than the wings which hold the sand, for if larger, the arbor could not be got out of the casting.

The ends of the core box should be made the same size as the flanges of the core arbor, then when the arbor is set in the box to make the core, these flanges will have a bearing on the ends, thus centering the arbor, and when the core is set in the mould the thickness will be equal all around. A half-core box is generally used for making the bottom in green sand, and the top portion is swept up; or a half-core box is set on the top, having the upper portion cut out so as to ram the sand through, as shown at *Y*. Lighter pipes, as a general thing, are made when green sand cores are used, than when dry sand cores are made in halves and pasted together. Of a lot of pipes made with dry pasted cores, the number that will have true round holes in them will be very small while; a lot made with green sand cores, made with a top and bottom box as described, will be found to have holes of the same diameter, and round.*

* Further valuable information upon this subject of "arbors" and green sand cores will be found on p. 140, Vol. II.

MAKING HAY ROPE LOAM CORES.

HAY, or straw, is wound around a core barrel for the purpose of venting the core, and allowing the barrel to be removed. When iron is poured around a hay rope core, it burns or chars the rope before the casting gets cold, thereby releasing the barrel so that it may be drawn out. With very large castings it is necessary to hoist the barrel out as soon as the iron is cool enough, for if left in till the casting is entirely cold the contraction of the casting will make it a hard job to get the barrel out. The rope also assists in holding the loam from dropping off the barrel.

In starting to put on the rope, tie the end with wire passed through two of the vent holes, and when the barrel is being revolved the moulder should keep the rope as tight as it will stand without breaking, so as to take all the stretch out of it. If the rope is put on slack, it will be impossible to sweep up a true, solid core.

When a core is large in diameter, it is best to have the rope pounded with a wooden maul as it is wound on the barrel. This will help to take all the stretch out of it. Should the rope break at any time while putting it on, take the end and let the barrel be turned back a little, thin out the broken ends for about a foot, twist them around each other and pound the splice so that it will not be any larger than the rest of the rope.

To fasten or secure the end of the rope when the barrel is covered, drive some nails through it and into the strands next to it. In putting the rope on the barrel, the barrel should be turned in the same direction it is to be turned when the loam is swept on.

When there is an offset or shoulder to be made on a core, or should the barrel be so small that one thickness of rope will leave too much loam, there can be another thickness of rope put on over the first. When doing this it is best to rub on some loam over the first layer of rope, so as to make a solid bed for the second layer. The first coat of loam that is rubbed on the rope should be very clayey and tough, and made so that it will work in between the joints and stick to the rope. If this first coat is not made to fill up all the joints and holes, the result will be a swollen or uneven casting.

It is also necessary to take a brick or block of wood and press it hard against the loam while the barrel is being turned, and if not sure that all the cavities are filled, it is a good plan to take a $\frac{1}{4}$ inch or $\frac{3}{8}$ inch round iron rod and press it between the joints while the barrel is being revolved, and then fill up the crevices with loam. After this, take the brick or block to lay all the loose hay flat, and make a solid surface for the second coat of loam, which should be put on so as to leave about $\frac{3}{8}$ inch for the finishing coat. The second coat should be used as stiff as it can be worked, to prevent bagging.

It is sometimes best to use the sweep in putting on this second coat, so as to make an even surface for the finishing coat. As a general thing, the finishing coat cannot be put on until after the barrel has been run into the oven and dried.

When dry and hot take it out of the oven, set it on the horses, as shown, and rough up the core with a brick so as to break the skin, to take the smoke black off. Take a brush and wet the surface slightly just before putting on the finishing coat. This coat should be put through a No.

8 sieve, mixed thoroughly, and enough put on the board or sweep to complete the core. When all is ready, take both hands and rub on the loam while the helper is turning the barrel. The barrel should never be turned fast. When turned once around, make the balance of the loam thinner and go around again. As the circle is completed, pull back the board endwise while the barrel is yet in motion. By doing this no mark will be left on the core.

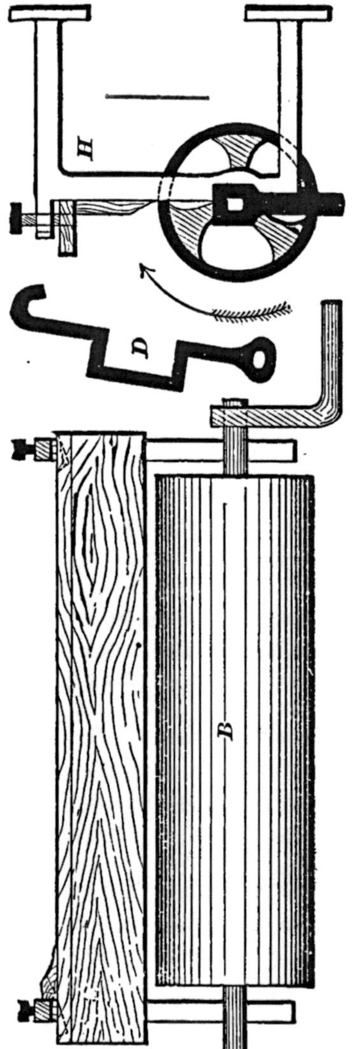

HAY ROPE CORE.

If you can finish the core in two revolutions, it will be smoother than if it takes four or five revolutions to do it. If there should be any rough places on the surface, instead of using a trowel, use a smooth, hard-wood block to smooth them with, dampening them with water and loam, and rubbing with the block till smooth and level.

If the core is large in diameter, it is best to keep the barrel turning slowly until the loam is set enough so it will not sag. If the barrel is stopped when the board is taken away, the core is apt to be out of round. In turning the barrel, never turn up against the sharp edge of the sweep, but towards the beveled portion, as shown. The thicker the casting, the more body or thickness of loam there should be over the ropes, especially if the iron is poured hot. When there is not enough loam over the ropes, the hot iron heats through the loam and burns the rope while the iron is yet liquid, and the iron will strain into the soft open places, and the casting will have lumps on it.

To make a casting one inch thick, the core barrel should be about three inches smaller than the finished size of the core. This will allow $1\frac{1}{2}$ inches on each side of the barrel for rope and loam. The size of rope should be about $\frac{3}{4}$ of an inch, which would leave $\frac{3}{4}$ of an inch for loam, $\frac{3}{8}$ of an inch of which should be left for the finishing coat. For castings about 2 inches thick, 1 inch thickness of loam will be safe.

During the war a firm in this country cast heavy cannons by coring them out, to save some of the boring and make a stronger cannon. The cores were swept up on a barrel, and, instead of using hay rope, they used ropes made of hemp. There were coils of small water-pipes in the core barrel, and cold water was kept running in them to keep the core barrel cool.

To make an even, strong hay or straw rope requires some practice, and the longest hay should be selected to make it hold together.*

The cut *D* shows a simple rope twister. Some use rope twisters made on the plan of a carpenter's bit-stock, which makes a handy tool, and *B* shows the barrel mounted on horses, with the screws for gauging the diameter of the core. *H* is an end view of the same. All core barrels should be well supplied with vent holes. Under 8-inch diameter they can be made of wrought-iron tubing. For small cores the hay can be put on the barrels without making it into ropes, as only a thin laying of hay is required. The same mixtures of loam that is used for brick loam work will generally do for this class of cores.

---

* For strengthening straw or hay ropes, a good plan is to twist the straw or hay around a tough string or twine.

BLACKING AND SLEEKING LOAM AND DRY SAND MOULDS.

THE poor quality of the blacking is generally the excuse made by many moulders for scabbing and for poorlv peeled castings. Sometimes such excuses are just, but in a great many cases the moulder who uses the blacking is the only one that is to blame. There are very few moulders that know how to mix blacking correctly, and sleek or finish a mould properly. At the present time the peeling of castings does not depend so much on the mixing of the blacking as it did fifteen or twenty years ago. In those days, when we bought blacking, we generally received it unmixed with resin, soap-stone, clay, black-leads, and minerals, etc. It was sold as ground, and free from the hard coal, coke, black-lead, soap-stone, or charcoal. When we ordered heavy blacking, we received a barrel of pure ground Lehigh; to this most every moulder had his own secret percentage of black-lead and charcoal, that he would mix with the Lehigh, when mixing his blacking in some unobserved place. At present we have only to go to the prepared barrel of blacking, and, as we do not know how much lead, charcoal, or anything else there may be in it, we take it from the barrel just as it is, and mix it.* In those days an experienced loam moulder could tell at sight of a newly opened barrel of blacking whether it was good or not; but now blackings are so mixed it is a hard matter to tell what it is until we try it. There is a way we can get some idea of the merits of blacking before that we put it upon our moulds. When

* A valuable chapter upon the "elements and manufacture of foundry facing" is seen on p. 208, Vol. II.

mixing up blacking, before it is thin enough to use, take a small ball of it, and dry it in the oven, and when dry see if it can be rubbed so as to make a dust easily; or, when the blacking is mixed up in good order, take a small core and black it over with it; when this core is dry, try to rub off the blacking with the hand, and then if it does not rub off easily, and seems to have a firmness about it, the blacking is generally satisfactory as far as the manufacture is concerned. There is such a thing as having the blacking mixed too strong, so as to make a poor mixture of blacking appear firm and solid when upon the mould; but when the casting comes out, it is blackened, scabbed, or the casting does not peel well. Ingredients can be used to wet and mix a blacking having no body in it, and yet it will appear very firm and strong when on the mould; but a trial test of blacking should be made by mixing it with a mixture of weak molasses or clay, water or beer, in order to decide upon its merits before using it. When a blacking can be brushed or rubbed off from the surface of moulds no one need expect to see the casting peel very well. When a blacking is so hard that we cannot scratch its surface so as to raise any dust, it is then mixed too strong, and it is very apt to scab or boil off when the iron comes in contact with it. Strong blacking is a good deal like the surface of a green sand mould that is made too hard, and it will cause trouble.

Many moulders think that the thicker a casting is the more blacking should be put on it. When $\frac{1}{16}''$ of thickness will not peel a heavy solid casting, it is generally safe to conclude the blacking has not been made and mixed properly; if $\frac{1}{16}''$ thickness of blacking will not peel a casting, the thickness of $\frac{1}{8}''$ will not do it. When blacking is put on thicker than $\frac{1}{16}''$, it causes the surface of a mould or blacking to generally flake off in spots, and the iron when it comes in contact with the blacking causes a gas; the blacking being so thick the gas cannot escape through to the loam or dry sand surface, and as it must free itself in some way, it will start and push out the face coat of blacking, and pass up through the iron.

When a casting commences to be less than one inch in thickness, then the blacking should be thinner upon the surface of the mould, especially towards the upper end.

When the casting is run altogether from the bottom of the mould, too much blacking on a mould for a thin casting acts as too strong a green sand facing on a thin casting; it will make the casting all cold shut. To properly put the blacking upon a mould in order to make a smooth-skinned casting is very important. The thickness of blacking should depend upon the condition of the surface of the mould. Rammed up dry sand moulds are generally about the same dampness when they are finished, but with loam moulds it is different; we sometimes do not get the blacking on the surface until it has become very hard or dry. When the surface of a mould is dry or hard, the first coat of blacking should be a thin one, the drier the surface the thinner the first coat of blacking should be, in order to have it soak in and adhere firmly to the surface. In putting on this first coat the brush should be rubbed up and down, and from one side to the other, as oftentimes only once passing the brush over the surface will not make the blacking surely work into the hard loam surface: the thickness of a second coat of blacking should depend upon how stiff or dry the first coat has become: if it is hard or dry, then the second coat should not be much thicker then the first was, and, of course, the thinner the coats of blacking are, the more coats must be put on when loam mould surfaces are dry and hard. It is best to black and finish one piece or section at a time, and after the first coat of blacking is on, the following coats should be put on before the under one has gotten too hard and dry.

We are very often forced to black loam moulds or swept up rolls while the loam or surface of the mould has hardly become stiff or dry enough to absorb the blacking. In blacking such green moulds we cannot use the blacking as thin as when blacking a hard surface, but it must be used thicker, in order to get body enough, and in such cases, when there are two coats required, we must be careful, lest in putting on the second coat we will take off nearly as much as we put on, in which case, when the casting comes out, we will wonder why it is that the coating does not peel better. It is always best to black a mould, if circumstances will allow, when the mould is just damp enough to soak up the blacking, so as to be sleekable about five minutes after it is put on, and also to have the blacking stay damp long enough to sleek the mould in good style, without having to bear on too hard with your sleeking tools to do it, since bearing hard upon tools when sleeking a mould is very injurious; for it not only compresses and closes up the pores of the blacking, but it also has a tendency to start it from the surface of the mould, the effect of which is not seen until the casting comes out, having some scabs upon it.

The less sleeking done in order to finish a mould the better. It is a good plan to lightly sleek once over the mould while the blacking is soft and damp. This will smooth down and fill up the hollows, and then to come back to your starting-point, by which time the blacking may be stiff enough to allow the finishing of the mould in good shape. Sometimes, when the blacking is very soft, the mould may have to be sleeked over three times before it has a good finish. A well-finished mould is one on which no trowel or any tool marks are seen, and also having all the parts sleeked smooth and the shape the pattern demands. If a square corner is required, see that it is made square, and not all filled up with lumps of blacking; or if a deep flange is needed, see that there are no streaks of blacking running down its sides, so as to make the casting look as if a lot of worms had been traveling over its surface, eating grooves in it as they went. Sometimes, when it is not easy to get at some crooked parts, in order to sleek them, many use a fine camel's-hair brush and some thin blacking for going over the surface. In fact, many moulders make a practice of doing this over all sleeked moulds, and it is a good way to do, when you want to quickly finish a mould, or hide any rough finish or tool marks. There is one redeeming quality about thus going over the surface of a mould. It will help to fasten down any spots or places that may have started from improper sleeking.

A mould when in process of blacking should have the blacking brushed or put on with a swab as smooth and even as possible, and not have it daubed on in any style, knocking off the edges and corners, and lifting up the surface sand. A mould blackened in this way is sickening to look upon. Time taken in order to blacken properly will be more than fully saved in the finishing—also will prepare a mould so as to be finished in good style, which it is impossible to do with a mould roughly blackened; and the attempt would only take twice as long as if the mould had been blackened smoothly and even.

In using a trowel or any tools to sleek or finish a blackened mould, the whole flat surface of them should never be used as a moulder does when sleeking a green sand mould. When too much surface is allowed to press or to be moved upon the surface of blacking, it will generally stick to the tools. To properly sleek blacking, the movements must be lively, and as little of the surface of the tools as possible be used; and also *never sleek twice where once should do.*

A dry sand mould is worse to finish, so far as the sticking of the blacking to the tools is concerned, than a loam mould. Sometimes, when finishing either of them, if the blacking has become dry, it will be started in an inexplicable manner, and cause the casting to be scabbed. The trowel should be slightly elevated or tipped up, so as to have only a small surface of the lowest portion touching, and if the blacking has become too hard for easily finishing, it is a good plan to dip the tools into water. This will help the blacking to sleek easier, and prevent its being started. When the blacking is soft, rub the tools with a good oily rag, which assists in cases where there is danger of the blacking sticking when being sleeked. Often in sleeking there are air bubbles formed under the skin of soft blacking, caused by too much sleeking, and which must be disposed of before a mould can be well finished. To do this, the air bubbles should be pricked with a pin or sharp vent wire.

An article that has been lately introduced, called plumbago, silver lead, or sometimes flake lead, is growing into great favor with moulders, as it is a great help not only in peeling the casting, but also permits faster and better finishing the mould. This lead, when of the right kind, is dusted by the hand over the surface of the blacking, and to give some idea to moulders of its merits that have never used any, it will be sufficient to say, that after it has been dusted on over the wet blacking, the flat of the hand can be rubbed over the wet or damp blacking, and there will not any stick to it. In using tools, they slide easily over the surface without any danger of the blacking sticking to them, and the sleeking of a mould is made a simple affair by its use. Blackening of moulds dry is a plan that is often practiced. There is less danger of a mould's scabbing when blackened dry than when green, since there is no sleeking done, and the blacking can be used thinner. The thinness of the blacking will depend upon the heat of the mould to be blackened. The hotter the mould is, the thinner should the blacking be. Moulds should never be blackened when they are so hot as to make the blacking blister. It generally takes from one to two coats more to blacken a mould when dry than when it is green, because the coats must be used thinner. To properly blacken moulds, either green or dry, will always require a mechanical judgment, and whenever there is any trouble with blacking not peeling or casting as it should let us investigate, to see if the trouble is not with ourselves before we commence to blame the blacking manufacturer.

IRON CASINGS FOR MOULDING POTS IN LOAM.

THE use of the flask or iron casing, as shown in the sketch, will be something new to many loam moulders. By this plan, instead of rubbing the loam on to bricks, it is rubbed on to iron. The pots made in these casings are used in a wire factory for heating wire.

In the morning, when the casings are hoisted out and when they are hot, the first coat of loam is rubbed on to them, and is about $\frac{3}{8}''$ in thickness. If the casings are not hot enough to dry the loam, they are run into the oven, and when dry and hot, are pulled out and lowered down on the shallow bottom and clamped.

There can be a thin sheet-iron ring placed between the joints, to project out to the face of the sweep to support the loam, and make a level joint.

After the center spindle is set into its bearing at the bottom, and secured at the top by the arm, as shown, the loam is rubbed on, the sweep passed around, and when this coat is stiff enough the finishing coat is put on and swept off smoothly.

The cut shows a sweep only half the length of the flask, that is, coming up to where the flask is jointed in the middle. They are made so, in order to make pots small in diameter, that would not admit of a man standing up in them to sweep them up. The lower section is swept up and finished, the top section put on, a second sweep is screwed on above the lower one, and the top section swept up.

Were the pots large enough to admit of a man working inside them, the casing could be made without any joint at the middle, and the sweep made the whole length.

The thickness of loam used on the surface of casings is 1″ at the bottom, tapering up to ¾″ at the top. This taper is to allow the casting to be hoisted out easily. The small sweep at the bottom is for making the bottom of the pot. This is not swept up until the rest is formed and hoisted off; then the top arm is fastened on at the lower bearing. These bearings are turned on a square shaft, which is better for fastening the sweeps to than a round one. The bottom is rammed up with dry sand. In closing the two parts together when dry, the joint must be secured so as not to leave a fin, which would prevent the casting from being hoisted off the casing. To insure this, it is necessary to go down into the mould and daub up the opening with blacking. To moulders that have never used casing for loam work, this plan would seem dangerous; but having worked with this rigging myself, and knowing that splendid castings can be made in a very short time by its use, I would recommend casings for castings of a similar character when there is a large number to make. For a few pieces it would not pay, as the rigging is expensive to make.

The main point in making such a rigging is to have plenty of vent holes in the flask or casing. The holes should not be over ½″, as the pressure of the metal would be apt to burst through them if larger. It is better to have the holes the largest on the inside. The first coat of loam that goes on should be as open in texture as possible.

These pots could be swept up flat ways, as well as in the way shown, by having the flask split in halves like a roll flask. A wooden frame, the size and shape of the cast-

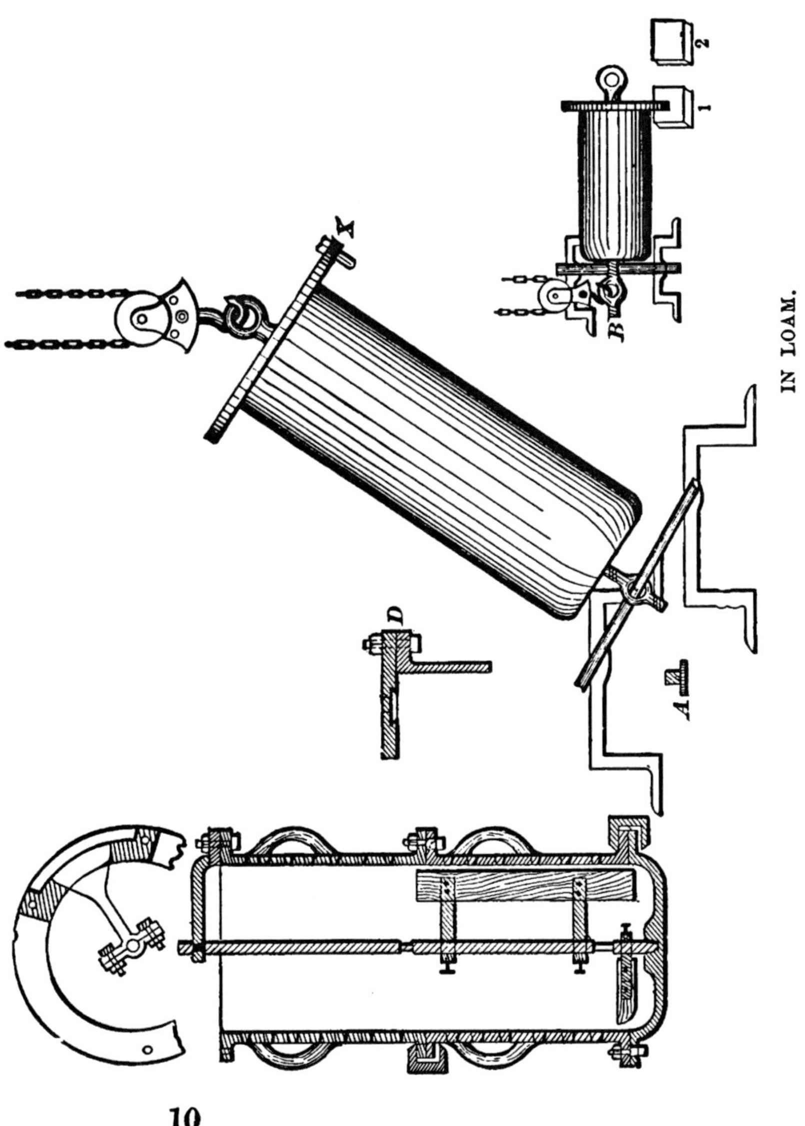

IN LOAM.

ing, should be made to lay on the joint of the flask when sweeping up the mould, to make the edge of the joint square and level.

A casting smaller than the one for which the casing was intended can be made by lining up the casing with brick. A flange or angle-iron, such as is used for a cupola, can be put on for holding up the brick. When building up the bricks put cinders between the joints and at the back to carry off the vent. The size of the pots made in these castings was from 2 feet 6″ up to 4 feet in diameter, and in length about 7 feet. The thickness of the castings run from 1″ up to 1½″. In pouring them, the iron fell from the top. The flanges of the castings should be turned up in the lathe ; also the broad flange on the core barrel, as it is the flange bearing on the top flange of the casing that supports and holds up the core. When the two pins (one of them is shown at *X*) are in their holes, you can rely on the thickness being equal at the bottom.

The core is a hay rope loam core, and in the two cuts is shown the manner of turning it over so as not to injure it.

The small cut shows the core barrel as it is hoisted off the oven carriage. It has to be dried standing on end. The blocks, Nos. 1 and 2, are used for assisting in throwing the core over, and when it is down the bar is put under the screw and the hook hitched on the square *B*, which is also used for turning the core barrel when making the core. The bar is then put through the eye of the screw or hook, the crane hitched to the other end, and the barrel hoisted up on its end, as shown, so as to be lowered down into the pot. The little plug *A* has a screw cut on it the same as the hook, and when the hook is taken out, this plug is screwed into the hole, loamed over, and blacked. A small fire of shavings is built under it to dry it. This plug has a square hole in it for screwing it in and out. The core barrel is cast with a bottom on, full of prickers. The top flange is bolted on the core barrel, as shown at *D*. In this flange there is a dovetailed groove cast at the point, to which the iron comes, and this is filled with loam, so that when the iron comes up it strikes sand instead of iron.

In fitting up the core barrel, the hook and screw must be central, so that the barrel will turn true on them, and the broad flange at exact right angles with the center bearings. The core barrel should have plenty of vent holes in it, and be made 3″ smaller than the size of core, to allow 1½″ on each side for the hay rope and loam.

DRYING MOULDS.

A WELL-DRIED loam or dry sand mould is a very essential point in making a casting that shall be free from scabs. Some irregularity may be admissible in the mixing of the loam or blacking, but the mould should be thoroughly dried. When the water in a damp mould is heated, it is converted into steam; and steam, when confined, creates pressure. Iron, when poured into a mould, heats up the surface and interior portions, and this heat generates steam if moisture is present, and the mould is very rarely strong enough or close enough to hold the pressure, which increases until it forces an opening through which it can escape. This may be towards the surface away from the iron, but it is more likely to be in a direction towards and through the iron. The outside of a mould is generally encased by an iron flask, or held by a curbing, between which and the brick-work sand is rammed hard and compact, and, with the exception of through a few vent holes, it is almost impossible for steam to escape in this direction. Towards the face of the mould the brick-work is open, or, if it is a dry sand mould, the surface is generally more porous than the backing, so that the steam will generally escape, or be drawn through the surface of a mould before it will find its way through the outside. This is the main reason why a damp mould will cause a casting to scab.

It may be asked, Why does not a green sand mould scab? The sand is damp. True, the sand is damp, but there is a certain limit to this dampness, which, if overreached, will cause trouble.

The surface of a loam or dry sand mould is generally hard and close, compared to that of a green sand mould, thereby permitting the steam generated at the surface of the mould to escape through the sand until it is free ; but should the green sand be rammed too hard, then the steam cannot force its way through, and it will come up through the surface of the mould and pass up through the liquid iron, thereby making a scabby or bad casting. A green sand mould that is rammed too hard, and a loam or dry sand mould that is not dried, have very much the same effect on the casting.

Whether a loam mould is dry or not is very often guessed at. The moulder will say it looks dry, and that as it has been in the oven a long time it must be dry. It is not the length of time a mould has been in the oven, nor the looks of its surface, that can alway be depended on to indicate its quality of dryness. A mould that should be dried in two or three nights is often only half-dried, as the oven may not work well, or there may have been some neglect on the part of the watchman. The fire may have been very hot for a short time, thereby scorching or burning the surface of the mould, while the interior is not half dry.

There is a great deal to be done in the way of properly managing a fire so as to save fuel and dry a mould as it should be dried. The first fire should be a slow and easy one, so as not to blister or crack the surface of the mould, which is caused by the efforts of the steam—quickly raised under the surface—to escape. This steam meeting the resistance of the half-dried blacking, which is very much like a sheet of rubber, stretches and blows it up into hills, but has not sufficient pressure to burst through and escape. There are many who think that by keeping a slow fire all the time to dry a mould or cores they save fuel. In some cases this may be so, as when a mould has little body, so that one night's firing will dry it ; but when a mould has a large body, after the first easy firing, in my opinion, there will be more fuel saved by keeping a good steady fire than by keeping a slow one. A slow fire will drive the heat in for about 8 inches, after which the further drying will be very slow.

With some moulds or cores this slow firing might be kept up for a week, and yet the interior not be dry. Whereas if the fire had been hotter the heat would have been forced into the interior and the steam and dampness expelled with probably two-thirds less fuel.

I have seen large cores put into an oven and orders given for a slow fire for fear of burning them, and after there had been fuel enough used to dry two sets of such cores, the boss would get disgusted because they were not dry, and give orders for a very hot fire, at the same time looking at the cores as if to say, "We will see who is to be boss." When the cores came out of the oven in a burnt condition, one could imagine them as saying, "Well, Mr. Boss, if you had used better judgment we would all have been well dried long ago, and not burnt either."

In the making of large body moulds or cores there should, if possible, be openings made from the center to the outside to assist the steam in escaping from the interior; and also, when possible, the center portion should be filled up with coke or cinders as much as can be safely done. The more coke or cinders the less sand and firing will be needed.

Plenty of venting in moulds or cores is also a great assistance in drying.

It is generally easy to tell when a dry sand mould or core is dry, but with loam moulds it is not so easy. Very few loam moulds are made, but the following plan could be adopted for determining if they are dry. Let the moulder, when building the bottom part of his mould, make an opening that will allow the inserting of a wet brick, or a lump

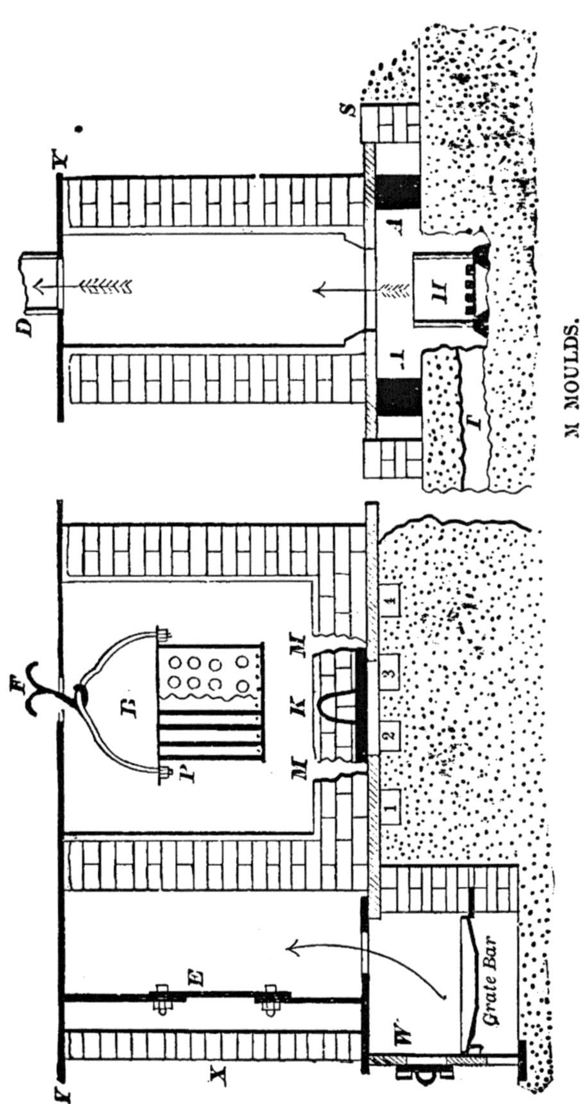

M MOULDS.

of wet loam, and on the outside of this let the opening be closed with temporary brick-work and mud. When he thinks his mould is dry, he can pull out the temporary brick-work and see the condition of the inserted brick or lump of loam. If this part of the mould has been placed away from the fire, and this brick or loam, when broken, is dry, he can generally depend on the mould being dry.

The cuts shown are for illustrating some of the ways of drying loam moulds that are too large to be dried in the ovens or too heavy for the crane to lift, making it necessary that they be dried in a pit or on the shop floor. *B* is a fire-basket, sometimes made in the form of an open grate-frame work all around the sides, as shown at *P*, and sometimes of boiler iron, drilled full of holes, as shown. For bottom grate bars in both styles, wrought-iron rods are generally used.

The baskets are made round or square in form, according to the shape of moulds they are to be used in. The width and height will depend upon the dimensions of the mould. There should be at least 18″ of space between the surface of the mould and the fire-basket, to prevent burning the surface of the mould before it gets thoroughly dried.

Sometimes, instead of using one large basket, three or four smaller ones are used, in order to better distribute the heat.

When the mould has a bottom in it, like the one shown, the baskets are generally hung by having the hook *F* held up by a crane or a strong bar. When a mould has no bottom in it, the basket can be let down so as to rest on bottom bearings. For moulds of this class, it is best, when possible, to have them hoisted up so as to have the bottom part of the mould about on a level with the top of the fire-basket, or have a hole dug so as to allow the baskets to get below the bottom, the better to dry the lower part.

When the inside of a mould is too small to admit of a

fire-basket being placed in it, a temporary fire-place is made adjacent to the bottom of the mould, and the heat made to pass up through the inside of the mould, by having the outer opening or space closed up with brick-work or sand, as shown at *S*. The fire-place shown at *H* can be made in the form of a basket, and placed directly under the mould or core. This basket can be pulled out to clean and renew the fire. Or there can be a temporary fire-place, as shown at *W*, built up outside of the core or mould, and the heat conducted through a channel to get to the inside of the mould.

To confine the heat, the moulds are generally covered over with sheet or boiler iron plates, as shown at *Y*, *Y*. *D*, is a stove-pipe, to carry off the smoke and create a draft. *E* is a sheet-iron curbing for retaining the heat. *X* is a brick wall for the same purpose. Either the wall or the curbing will answer the purpose.

The combined fires are only generally needed when there is over an 8″ wall to be dried, in which case a fire in *W*, so as to heat up the outside of the mould, is combined with a fire on the inside of the mould, and also channels, as 1, 2, 3, and 4, connected with the fire *W*, to carry the heat underneath the bottom plate, which should have plenty of holes in it. The two fires thus combined will thoroughly dry a mould. These channels can be formed by using brick, or rough gutters can be made in the sand, either of which, if desired, can be filled up with sand after the mould is dried. The heat could be got under the bottom by having the plate raised on iron blocking, as shown at *A*, *A*.

It is always the bottom portions of such moulds that are the hardest to dry, especially so if the mould is built up in a pit. In such cases it is a good plan to have, when possible, a large hole cast in the bottom plate, so that when the bottom is being bricked up there will be a part of the mould left open.

Then below this opening in the mould let there be a small pit dug with a channel. *T* is a pipe laid to the outside of the mould to admit air to the pit, creating a draft. Then, with a fire-basket lowered down through the mould into the pit, we should have a fire below the bottom of the mould the same as shown at *H*. After the mould is thoroughly dried by the combination of fire-baskets *H* and *B*, the pit is filled up with sand, and a plate having built upon it bricks or core sand, and previously dried and still hot, is lowered down to fill up the opening, as shown at *K*. Between the two plates there should be a little soft loam to form a solid bearing. The open space *M, M*, is then filled up with a dry mixture of loam, and should the top surface not be even with the original surface, it is made so by filing off or building on. A thin sheet-iron plate, having on it a charcoal fire, is laid over so as to assist in drying out dampness.

Sometimes when building the bottom of a loam mould that is very thick, it is best to partially dry the bottom brick-work before the upright portions of the mould are made; which can be done by having the plate raised up and a wood fire underneath and a charcoal fire on top. After this the bottom can be permanently set where wanted.

The best kind of fuel to use in the fire-baskets will depend on the draft. Charcoal requires the least air, gas coke more, and soft and hard coal and coke the most.

The moulder must use his own judgment as to the best plan to be adopted for drying any particular mould, as there are hardly two moulds that the same drying arrangements should be used for; but it is hoped that from some of the different plans given there can be found one that can be turned to answer his purpose.

CHAPLETS AND THEIR USE.

In making castings that require the use of chaplets, the moulder is frequently annoyed by complaints about blow-holes. If asked what caused the blow-holes, he would be likely to say that the chaplets must have been rusty, of which there can be no question. To know what this rust is, and its chemical action when surrounded with hot iron, should be of as much interest to engineers and machinists as it is to the moulder. Any one employed in a foundry knows—some to their sorrow—that to take a rusty rod and quickly push it into a ladle of melted iron will cause the hot iron to fly in all directions. This is caused more from the dampness than from anything else, as all rusty iron is more or less damp, and hence, when plunged in the hot iron steam is instantly generated, which scatters the iron in its efforts to escape.

To demonstrate this, take a rusty rod, heat it enough to dry up all the moisture, and then put it into a ladle of iron. The iron will boil around it more or less, but will not fly over the foundry as it would if the rod was not dry.

There are two things to be contended against in the effort to keep melted iron from blowing or boiling when enclosing rusty iron. The first is steam, and the second is carbonic oxide gas. This gas is formed by carbon in the hot iron combining with oxygen.

Take a piece of polished iron, and let it get damp from the moisture of the air—or otherwise dampen it—and it soon becomes rusty, because of the affinity of iron for oxygen when combined with water. Under certain conditions polished iron can be kept from collecting rust or oxygen; as by keeping it no colder than the temperature of the air, and keeping the air dry. Take the iron from a cold room into a warm one, and it will not be long before rust will collect on it. This is caused by the cold iron condensing whatever moisture there may be in the air.

To illustrate how much gas is formed in a mould when melted iron comes in contact with rusty iron, I cut off a piece of ½″ round iron one foot long, and had it weighed on a pair of fine scales. I then took the rod and heated it red hot, so as to burn off all the rust, after which the rod was reweighed and found to weigh sixty grains less.

To know how much gas this sixty grains of rust or oxide would form, I submitted the matter to a chemist, Mr. L. H. Witte, who found that sixty grains of rust in melted iron would make thirty-one grains of carbonic oxide gas, which at 2,800 degrees of heat (the melting point of iron), and a pressure of one atmosphere, would occupy about six hundred cubic inches of space. The volume, or space, which gas occupies depends on the pressure. If a moulder sets rusty chaplets, the damage will be proportioned to the temperature and pressure of the iron around them. It is very seldom that chaplets in common pipe and similar castings should have blow-holes around them on the side cast down. It is on the top or cope part, where there is very little pressure, that the blow-holes are found. The same may be said of cylinders or other castings where chaplets are used.

The question might be asked, Why is it, that where the greatest pressure is, the gas escapes the easiest and without causing blow-holes? The parts of a mould where the greatest pressure is are usually the first to be filled, and the iron is hotter and cleaner than at the top of the mould. Should the chaplets at the bottom cause the iron to blow or boil, the gas will escape upward through the iron, and come out of the mould at the runners or feeders. The iron being hot, the pressure will not allow any holes or cavities to exist; but should the iron boil or blow around chaplets in the upper sections of a mould, it will generally leave blow-holes in the casting, because of the iron being dull, or having no life in it, so that the gas cannot escape through it, but stays around the chaplets. The size of the cavities will depend on the amount of gas formed that cannot escape.

Chaplets are very often kept in moulds for two or three days before the mould is cast. In such cases they are very apt to corrode or get rusty, especially if the mould is a green sand one.

The moulder may paint or varnish the chaplets, to proteet the iron from getting rusty, but the paint or varnish will sometimes create more gas than the rust or oxide would. Again, the paint may be of such a nature as to proteet the iron chaplet from rusting, but hold moisture itself, and when the melted iron surrounds the wet chaplet it forms a cushion of steam around it, and the blow-holes are formed, the same as from the gas caused from the rust.

About the best thing to prevent chaplets from blowing, or boiling the iron around them, is to have all the rust burnt off and have them tinned over, which can be done to advantage for a standard class of work. The affinity of tin for iron makes the iron hotter. Pieces of tin are often thrown into ladles of iron to make the iron more fluid. The tin, beside making the iron around the chaplets hotter, so as to give any gas that may be formed a better chance to escape, also protects them from collecting moisture and getting rusty.

For castings where it is essential that the upper section

shall be sound, it is well to use what is called a loam chaplet. This is made by taking solid iron, wrought or cast, and daubing the surface exposed to the melted iron with a thin coat of loam.* This will leave a clean hole in the casting, which the machinist will have to tap and plug up, but when the casting is put to the test, there will be no danger of blow-holes around the chaplets. In using such chaplets pieces of iron can be built up on top of the core arbors so as to come even with the face of the core, and have the chaplets rest on iron instead of on sand. By this method fewer chaplets will be required to hold down a core. The fewer the chaplets used the better and stronger the casting.

Red lead mixed with turpentine is one of the best paints for chaplets. Chalk, coal tar, oil, asphaltum, etc., which are often used on chaplets, are not so reliable. In some shops cast and wrought chaplets are used very extensively. The cast-iron ones are the best to use on castings that require to be finished, as the melted iron adheres to them better than to wrought-iron ones. In some cases where castings are finished, the chaplets cannot be seen.

Cast-iron chaplets can be made of any shape or size, and used in castings from ½″ up to 3″ thick, but care must be taken not to set them where the gates will cause the iron to run against them as they melt very readily.

The cuts Nos. 1, 3, 4, 5, 6, and 7 show a class of wrought and cast iron chaplets that are very handy for most classes of work. No. 1 is a cast-iron chaplet that can be made very readily from ⅜″ to 1″ diameter, and of any length required. They are made by ramming up a deep flask having a level joint, and after the cope is off, bedding in the heads and driving down the long stem any length wanted. They should be notched with a chisel on each side before knocking them off, as they are apt to break below the surface of the casting when roughly knocked off.

---

* Often by having a thread cut on the stem of chaplets and painted with red lead, excellent solid castings can be produced.

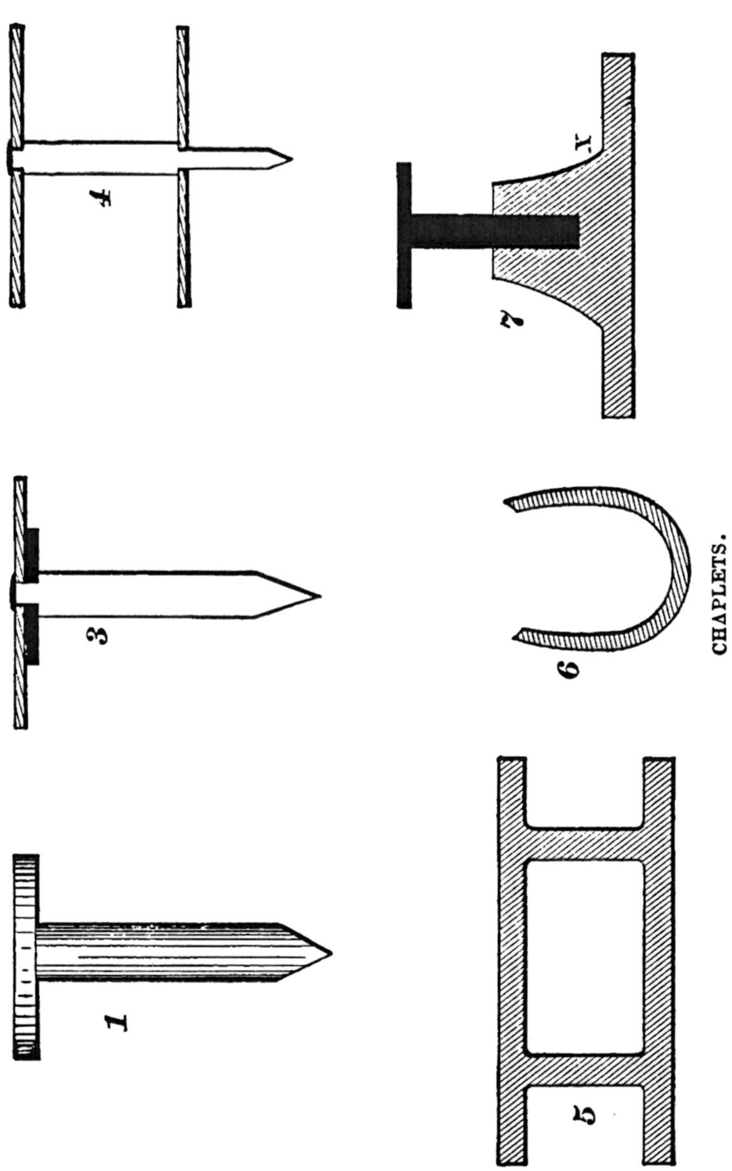

CHAPLETS.

No. 5 is a double-headed chaplet, used particularly on loam and dry sand work. A pattern will be required for each size wanted.

No. 7 shows a cast-iron chaplet and stand which is very handy for loam, dry, or green sand moulds. The face of the stand *X* is set against the pattern and rammed up, or built up in the brick-work. The only objection to using this stand is, that it will chill the casting, for which reason the stand should not be set on castings that require to be hard iron. These stands are better if cast solid, and the holes for holding the chaplet drilled out. When setting the chaplets, if they are too long, break off a piece, and if too short, fill up the holes with sand. Iron flasks for special jobs often have holes drilled in the bars to hold chaplets, whereby much time and labor is saved.

No. 3 is a wrought-iron chaplet, having a large double head riveted to the stem. This is safer than having only a single head riveted on, especially for large cores that have a heavy lift under them.

Some blacksmiths can take a nut, and by putting a shoulder on the round stem, weld the nut on, making a head on a chaplet 3 or 4 inches broad, which is safer than a head riveted on over a small shoulder.

Chaplets that do not require very large heads can be made cheaply in a machine for heading bolts.

No. 4 is a double-headed wrought-iron chaplet, having a sharp stem, to be used on loam and dry sand work. The top head is riveted on, and the lower one is made to slide up on the stem to a shoulder, which is filed to make the chaplet the size wanted.

No. 6 is a spring chaplet made from hoop, sheet, or plate iron, bent with the grain of the iron. This is very handy for placing between cores where it would be hard to make a stiff chaplet stay. Sometimes these chaplets have their ends bent inwards so as to come in contact with each other, thereby making a stiffer spring chaplet.

As regards the size of iron for making chaplets, the moulder must use his own judgment, as different castings require chaplets of various sizes and strength. Chaplets are a very important feature in the manufacture of castings, and are always an eyesore to look on, as they disfigure castings more or less. A good moulder will use as few of them as possible.

In Vol. II., p. 178, is a chapter entitled "Improper Setting and Wedging of Chaplets," which, if read in connection with this, will be found to give the subject of chaplets a thorough and valuable presentation.

LEAVING RISERS OPEN OR CLOSED ON LOAM OR DRY SAND MOULDS.

Among loam moulders and shops in the practice of casting loam moulds, the question of whether the risers are to be left open or closed seems to have been established more from custom than from any thought upon the subject. The custom of one shop is to cast all loam moulds by having the risers open; another shop would not permit such a thing, and it has often been a matter of thought whether such customs did not prevail simply because it was the practice of other moulders. There is no doubt many moulders leave risers open or shut after careful thought and study upon the subject. In giving their views, it is possible some may differ; but if they do, it will result in their giving thought to the subject, and not acting blindly.

When iron is poured into a mould which has all the risers closed up tight, the air in the mould is compressed. Iron dropping into compressed air cannot drop with such a dead fall as it would if there was no compression; and iron running into iron, having a pressure of air upon it, cannot rise so fast as it would, if there were no pressure.

Compressed air in a mould will often prevent its scabbing and the surface gases from coming inwards. These facts seem to give good and sufficient reasons for drawing the following conclusions: when iron is poured into a loam mould from the bottom, it is very often best in thin castings to have the risers open. This will allow the iron to rise up more freely and faster. Whenever a mould is cast open, the area of a riser or risers should be large enough to permit the air to pass off freely and without a noise. It is often best, when the iron is on the dull side, to leave the risers open, especially in such castings as steam cylinders, etc., which have many cores in them. The cause of blow-holes in the upper portion of such castings has arisen from the dullness of the iron not giving collected gases or air a good chance to escape through it. When a mould is burnt very badly, it is better to keep the risers closed, as there will be a compression against its surface, instead of a laxity and rushing upwardness of blasts of air and gases. When pouring a mould by dropping the iron from the top, its fall and cutting actions will be made easier upon the moulds by having the risers closed ; for such castings as rolls, spindles, or cannon, risers heads are generally left open. It may sound odd to some moulders to read such an expression as the drawing down of loam or dry sand covering plates or copes ; but the writer has seen the cope surface of anvil block castings all covered with what the moulders called scabs. And to prevent them, they used different mixtures of sand or loam, and all to no purpose. They were then told if they would close up their feeding riser heads, so as to allow no air or gas to escape, the trouble would be stopped ; but the advice was laughed at, and it was not until they saw it practiced and the results obtained from it, that they believed loam and dry sand copes could be drawn down. It is not very long ago a certain foundry had some heavy fly-wheel to make, and the rim being covered over with a loam ring, the cope part would be all drawn down, so a remedy was sought for, and it was not until the risers or feeders were made air-tight, and the joint also, that good wheels were cast.

A flat cope surface of loam or dry sand, when exposed to the direct heat of a rising heavy body of iron, will be drawn down upon the same principle as green sand copes are drawn down, and any one who doubts the truth of this will be convinced sooner or later of its correctness.

RESERVOIRS AND LADLES FOR POURING HEAVY CASTINGS.

When pouring heavy castings there is usually a feeling of suspense and anxiety experienced by all interested. It is in the few moments that the moulds are filling with iron that the work of weeks, perhaps months, is tested. The least neglect or wrong-doing in the construction of the mould may make the pouring unsuccessful, thereby involving a loss of hundreds of dollars. The moulders that this class of castings can be trusted with are few. They must be long-headed, cautious *mechanics*. In some cases a man that is not a thorough mechanic may, when the work is planned and laid out for him by his foreman, be trusted with large responsible jobs, if he is a steady, thoughtful, and cautious man. When a mould is being poured, should anything go wrong, it is very rare that it can be remedied. There is but one trial for a mould, and during the pouring there is no such thing as waiting a while to fix the part that is wrong.

In the engravings is shown a ladle calculated to hold ten tons of iron ; also the construction of a reliable reservoir for receiving and holding large quantities of metal, until enough is melted to pour a large casting.

When melting iron for very heavy castings, cupolas and air furnaces are generally used. From the air furnaces iron spouts or troughs connect with the reservoirs, so that when all the iron that has been charged up is melted down and found to be of the right temperature, the furnace is tapped out and the iron let run into the reservoir. At the same time the cupolas will be in blast, and the melted iron conveyed from them to the reservoir in crane ladles, until by measurement there is found to be enough iron in the reservoir to pour the casting. Then the iron is let run from the reservoir into the mould.

There are several ways of constructing reservoirs for holding iron. Sometimes a lot of pig iron can be built up in a circle to form a green sand reservoir. The pig iron gives a backing to the green sand, preventing the reservoir from bursting. Under such green sand reservoirs it is best to have a coke bed, and, to form an outlet, loamed plates can be used, having the pig iron placed on each side so as to form a slide for the plate to work up and down in. The pigs must be protected from the melted iron by green sand. To form the bottom of the outlet, there should be a dry sand core used, to prevent any washing away when the iron runs out.

In making such reservoirs the sides are built up with considerable slant, so as to make the bottom smaller than the top. This gives strength to the bottom. A sheet-iron curbing would answer the same purpose as the pig iron, and would be better for green-sand reservoirs from 4 up to 8 feet mean diameter.

For more than 8 feet it is safer to make a reservoir as shown in the engraving, in which *E*, *E*, are iron plates bedded on a solid flooring; *P* is a boiler iron curbing, the plates being screwed together with bolts. Inside of this curbing is sand, rammed solid, and on the top of this solid foundation a stout cast-iron plate, *Y*, is bedded. To this plate *Y*, is bolted the reservoir curbing, as shown at *B*. For very deep reservoirs this angle iron should be one continuous ring all around the bottom of the curbing. It can be made of cast iron or of boiler iron. It is not necessary that

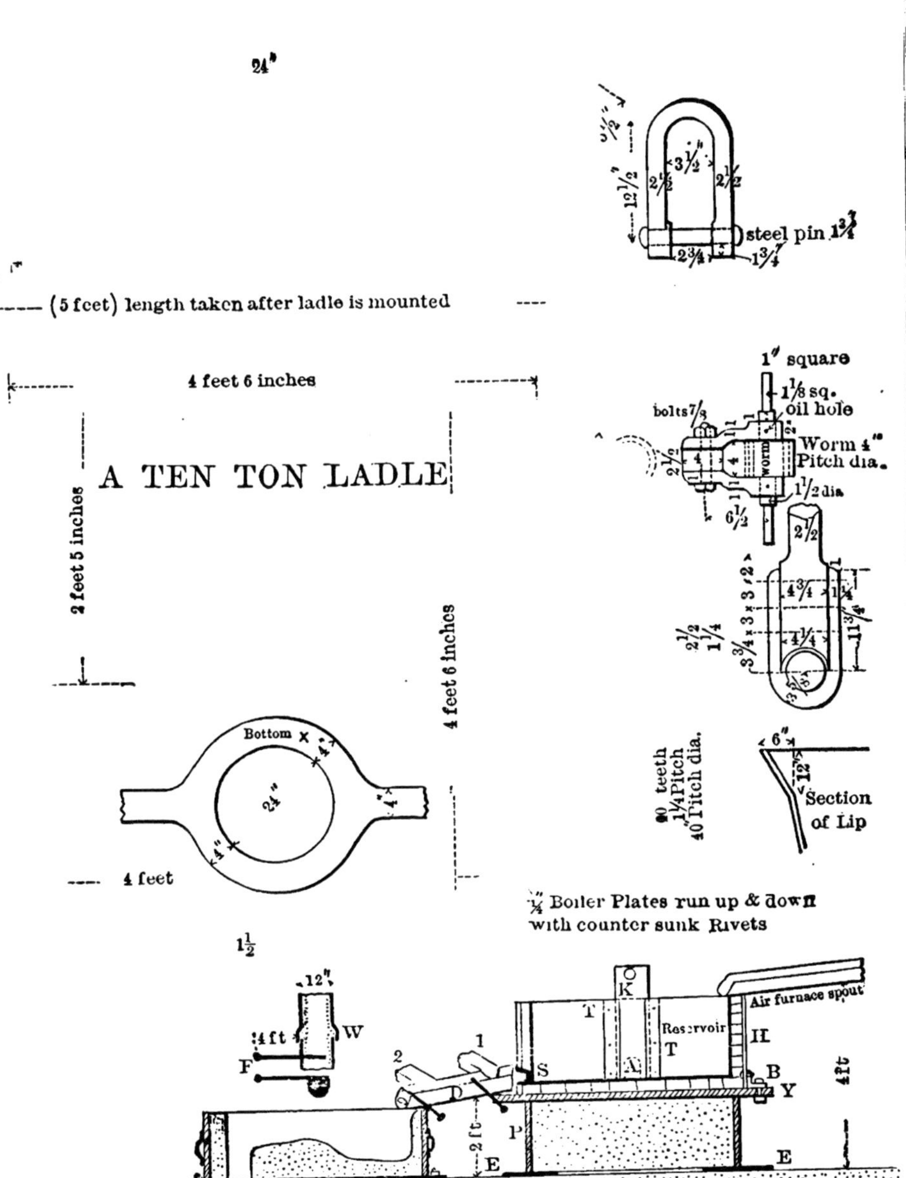

24"
(5 feet) length taken after ladle is mounted
4 feet 6 inches
A TEN TON LADLE
2 feet 5 inches
4 feet 6 inches
Bottom
24"
4 feet
steel pin
1" square
oil hole
bolts 7/8
Worm 4" Pitch dia.
Section of Lip
Boiler Plates run up & down with counter sunk Rivets
Air furnace spout
Reservoir

the bolts should be very close together. If the angle iron is heavy, bolts from 3 to 4 feet apart will answer.

For an outlet, a tapping hole, *S*, similar to the tapping hole in a cupola, may be made, the only difference being in the stopping used.

In stopping up a cupola tapping hole clay is generally used, and the tapping out is all done with bars; but in the tapping of a reservoir it is only once done, and if the tapping hole is entirely stopped with clay, it will get baked so hard as be likely to cause trouble.

About the best way is to fill up the tapping hole with sharp sand wet with clay wash, and then in front of the hole lay a piece or two of pig iron, so as to make sure of holding the pressure.

When all is ready to run the iron into the mould, remove the pieces of pig iron, and with a mason's trowel dig out the sharp sand (throwing it away so that it will not run into the casting), until the sand looks red-hot, which is a sign that there is not much thickness of sand left. Then, with a sharp bar, the iron can be tapped out without any danger of knocking in the breast or of making the tapping hole any larger than wanted—a danger that always exists when the tapping hole is stopped up with clay, or with any substance that will bake hard.

As regards the size of the tapping hole, it should be smaller than the runners that admit the iron into the mould, especially so when the casting is run from the bottom of the mould, for the iron in the reservoir has a head pressure to force it out, while the free flowing of the iron into the mould is retarded by the friction of the gates and runners. Also, as the iron rises up in the mould, the slower will the flow be.

Another plan for letting out the iron is shown at *K*, *T*, *T*. Here the outlet is made on the principle of a damper and slide. *T T* are slides that are secured to the curbing. *K* is a cast-iron damper, or plate, having prickers cast on one side, this side being daubed and finished up with loam, and dried. The damper is then set in its place, and whatever open space there is left between the slides and damper is filled up with loam or stiff blacking. This prevents any leakage of the iron. When all is ready the damper can be raised as wanted by a lever or with a crane. The iron will then flow out through the opening as marked by the dotted lines at *A*. Should the iron come out too fast the damper can be weighted down so as to shut up the opening *A*, as desired.

These reservoirs are sometimes used for the purpose of making sure of having the iron well mixed. In making very heavy castings, the iron is sometimes melted in a number of furnaces or cupolas, and the iron as taken out into separate ladles is seldom alike in quality. If a casting is poured from the ladles there will not be a uniformity of iron throughout the casting. If instead of this all the iron is first collected in one mass, there is a good chance that the grade will be uniform throughout, which for many castings is a very important consideration.*

Some foundries have large, deep tanks, made to hold from 10 up to 20 tons of iron. These are made of boiler iron, similar in shape to a crane ladle, but of simple and inexpensive construction. When there is a heavy casting to be poured, these tanks will be set upon solid blocking, and the iron will then be brought from the furnaces or cupolas in crane ladles and poured into the tanks. When the tanks are full, or have enough iron in them for the purpose, they are tapped from the bottom of the tank. These tanks are generally lined up with fire-brick, and are kept in some part of the shop where they can be hoisted by cranes and placed wherever wanted.

Sometimes these tanks have trunnions and under-straps

* Also, when thus collected in one mass, the temperature of the metal will be uniform, and can be better managed if desired to be cooled.

fastened to them, and instead of supporting them on blocking, the tank will be held up by having the trunnions rest on strong iron horses.

There are generally spouts or troughs used, as shown at *D*, to convey the iron into basins before it enters the mould. These troughs are generally made of cast iron, daubed up with loam and dried; when there is a long run required, the troughs are united, as shown at *W*, and the joints are daubed up and dried with hot irons or fire.

The basins used to receive the iron from the reservoirs or tanks should be made large, so as to give a good chance to regulate for fast or slow flowing of the iron. Such basins are the better and safer if made of dry sand or loam.

Pig beds are usually made to receive the overplus iron, of which there should always be some in order to insure against pouring the casting short.

Sometimes reservoirs or tanks are used in connection with crane ladles. In this case pig beds will seldom be required, for the iron in the tanks or reservoirs can be so calculated as to make it sure that it will be all needed, and then the iron in the ladles can be poured out until the mould is filled up, at which moment the pouring can be stopped and the iron left in the ladles can be used to pour some lighter casting with, thereby saving the cost of melting a lot of extra iron, and the labor to handle it twice, saying nothing about the mess a lot of iron makes when poured out on a foundry floor.

Sometimes castings are run directly from the air furnaces without using any reservoirs or tanks. The iron will run along through spouts or troughs into a basin, and from the basin into the mould. A branch trough is arranged, so that when the mould is full, by raising an iron shute the surplus iron is allowed to run into a pig bed.

Often there will be two or three castings poured directly

from one air furnace. One is first poured, and then an iron shute is raised, and the iron made to flow into the second mould, and so on. When arranging these branch troughs to pour two or three castings at one tapping, the main trough *D* must have more or less of a fall to it, according to the length of the run, and the first casting to be poured should be connected with the highest branch, No. 1. To pour the second casting, the upper shute, after the weights are taken off, is lifted up, and then the iron flows down into branch No. 2. The branch No. 1 is then closed up by using an iron stop, as shown at *F*, and by shoveling in some sand and putting in pieces of pig iron at the back. After all the moulds are full, the surplus iron is run into pig beds, as described.

The ten-ton ladle shown was copied from a tracing loaned by a friend, Mr. *John T. Stoney*, a moulder, and superintendent of large experience. The ladles thus made have given the best of satisfaction as regards durability and easy working. Measurements are given to assist any one who may want to build a first-class screw ladle for carrying up to ten tons of iron. For a fifteen-ton ladle it would be safer to have the parts enlarged. In building ladles, the top diameter and the depth are generally made the same. Taking this for a rule, the following figures will show the capacity and sizes of ladles usually wanted in a foundry·

CRANE LADLES.

Top Diam.	Bottom Diam.	Depth.
46″	40¼″	46″
41½″	38″	41½″
37″	32½″	37″
30″	26¼″	30″
24″	21″	24″
19″	16½″	19″
12½″	11″	12½″

HAND LADLES.

Top Diam.	Bottom Diam.	Depth.	Capacity.
11″	$9\frac{1}{4}$″	11″	230 lbs.
9″	$7\frac{3}{4}$″	9″	126 "

The foregoing figures represent inside measurements of ladles when lined or daubed up ; so that, if the ladle is to be lined up with fire-bricks or clay daubing, the thickness of the bricks or daubing to be used must be added to the diameters and depths given. To test the amount a ladle will hold, can be told by filling a ladle full of water, and multiplying its weight by seven (the approximate specific gravity of molten iron).

For lining up reservoirs that are only intended to be used once, common brick can be used, and on the surfaces exposed to the liquid iron a coat of loam rubbed on, as shown by the heavy dark line on the surface of the bricks. After the bricks and loam are well dried by fire, a good coat of blacking is applied to the surface of the loam.

To line up large crane-ladles, or tanks, fire-bricks are used, as shown in the cut of the ladle. The angle irons shown are for holding the brick or the clay daubing, although some foundrymen will not use them, thinking them more trouble than service.

For daubing eight down to four-ton ladles, fire-brick are generally used for the bottom, and on the sides a stiff clay daubing is used.

Below four tons, clay daubing is used on the bottom as well as on the sides. The thickness of clay on the bottom is from 1″ up to $2\frac{1}{2}$″, and on the sides the clay is thicker at the bottom than at the top, running from 2″ to 1″. For small hand ladles, the thinner the daubing can be used the handier the ladles will be.

It is not the thickness of the clay or daubing that is to be depended on so much as it is the being sure that all the cracks in the daubing are well closed up, and that the clay is of an equal thickness all around the ladle.

SCABBING OF GREEN SAND, DRY SAND, AND LOAM MOULDS.

Any section of the mould that is covered in four or five seconds with a body of iron about two inches thick can be rammed harder, and will require less venting, than if it took a longer time to get this body of iron over it. The thicker the body of iron, the more the air, steam, and gases are forced to escape downward through the vents and sand. Wherever a scab is seen on a casting, it is certain that the iron bubbled and boiled at that point when pouring.

There must be a bubble before there can be a scab. The bubble may be caused by hardness, closeness, or wet sand. The wet sand causes steam to be raised, and for release it will follow the direction of the least resistance. The same result will follow from hard ramming or closeness of sand. The air and gases cannot escape fast enough through the vent and sand, so they lift or escape through the surface of the mould and through the iron, causing bubbling and scabbing of the mould.

There are instances in casting where the lower part of the mould is filled quickly, and as the rising iron comes up, it has to cover over a large surface projection, or green sand cores, from which point upwards the casting is a solid body of iron; therefore the iron does not rise so fast in this section of the mould, thereby causing the top surface of a green sand core or projection to be slowly covered with the rising iron, which will cause such parts to scab very easy, if the greatest care and judgment are not used.

For such cases it is a good thing to mix some sharp sand, like lake or bank sand, in with the moulding sand, using this mixture as a facing sand. For the top surface of a projection or core, a mixture, such as one part of lake or bank sand, mixed with two of moulding sand, will allow the surface to be firmly rammed, and still be open enough to allow the rising iron to quietly lay on it, no matter how long it is before a pressure or body of iron is raised upon it. Projections or cores in a mould generally need to be rammed and rodded well, and are the parts that need the greatest care. It is also important to keep all risers and feeding heads closed air tight, as, when they are open on this class of work, the air rushes out, taking the pressure of the air and gases off the surface of the projection or cores. As the lower part of the mould fills up first, the gases and vent from it will be drawn by the escaping current of air through the risers. This combined current, in escaping, lifts or starts the surface of the projections, and when the iron comes up to it, the iron fills up all the vent holes and sets the mould blowing. A blowing mould from this cause is a dangerous one. In some cases it will not stop until all the iron is blown out of the mould. The above are a few of the many reasons for lost and scabby castings in green and sand moulds

As regards scabbing in loam and dry sand moulds, the first and greatest cause is in not having the mould well dried, as the steam generated in process of casting acts on this class of work in the same manner as it does on green sand work. The only difference is, that sometimes the loam or dry sand buckles, or is pushed out into the molten iron. This buckling is caused by confined steam, gases, air, or some kinds of close sand. The buckling in such cases is not like a green sand scab. When the casting is taken from the sand, a few blows with a hammer upon the scabbed spot will cause the lumps of sand to fly out, leaving holes in the casting. In the case of a green sand scab, the sand is generally found in some other part of the casting. In loam or dry sand there is not the amount of air or gases to be carried off by vents that there is in green sand. It is only in pockets, corners, and under flanges that it becomes necessary to carry off vents directly. With a mould well dried, and the loam, sand, and blacking mixed in the proper proportions, there is very little danger of scabs. Having had about an equal practice in the three branches, I can safely say, that green sand work, so far as scabbing is concerned, is the most difficult to contend with. Little does a looker-on know of the unseen injury to castings which a moulder can do by ramming the different parts of his mould too lightly or too heavily, or not venting it properly.

CONTRACTION AND CRACKING OF CASTINGS.

THE query of why a casting cracks, is generally looked upon as a conundrum ; at least the different answers to the question, and the different theories on the subject, would lead one to think so. There are very few things about a foundry that seems to be so little understood as the contraction of iron when cooling. Few appear to know whether there is any difference in the contraction of thin and thick castings, hard or soft iron. Is the contraction of iron gradual, or is it true that castings have little or no contraction vertically, as is thought to be the case by a writer on "Why don't Castings Shrink Vertically?" in the *American Machinist*, March 26, 1881, and signed "Moulder."

The writer of the article referred to will please excuse me for not noticing it before. My only excuse is that I was waiting to see if some one else would not answer it.

My opinion is that castings *do* contract vertically, which opinion is borne out by experience and by practical tests which I have made.

Not long since I had two patterns made, 54″ long by one inch square. These patterns were moulded in two flasks. One flask was suspended so as to cast the mould vertically, and the other was cast horizontally. When moulding the one that was to be cast vertically, care was taken to ram it so that there would be no straining on the sides, and to prevent the lower end from straining there was an iron chill rammed firmly up against the end of the pattern. These two flasks were cast out of the same small ladle of iron. If "Moulder" will try his theory by the above practical test, he will see whether it is not correct, for he will not be able to tell, so far as the contraction is concerned, which of the castings were cast vertically. This is as fair a test as could be made to determine whether castings contract vertically or not. The moulder may say he has measured castings that have been cast vertically, and found no contraction. This may be all true enough. There are lots of castings thus made which will measure longer than the pattern. So also many castings cast horizontally have been found to be larger than the pattern. Again the question could be asked, "What is the cause of this?" to answer which it can be said: "The straining of moulds."

A moulder or pattern maker that has had experience in making heavy rolls will, or should, always make the lower end of a swept mould, or the pattern, smaller than the end that is cast up. How much smaller will depend on the length and body of the casting. The difference will vary from $\frac{1}{16}$ up to $\frac{1}{4}$ of an inch.

I have seen a rolling mill boss come into a foundry with such fire in his eyes that the poor moulder has trembled all over when asked, "Why in the name of common sense he could not make the lower end of his rolls the same size as the upper end, and not have the wabbler so much too large that it would take a good chipper a week to chip it so as to have the coupling that goes easy on the top end fit the bottom end?"

That castings contract vertically has often been proved in making rolls. I have seen the upper wabblers cracked and pulled off from the neck of rolls, from no other cause than the contraction of the casting, and the carelessness of the moulder. That is, when feeding up his roll with hot iron to supply the shrinkage, he would let a flange form on top of his feeding head, so as to come out and rest on the iron flask or boxes that are used for forming the feeding head and when the cooling crust commenced to contract it was held up by this flange. Of course it formed a crack or breakage in the neck, or wabblers, because the half molten metal was not strong enough to lift up the whole roll from the bottom of the mould.

In moulding castings that are cast vertically there is generally more or less straining of the bottom portion of the mould, which in many cases cannot be avoided. Loam and dry sand moulds are strained more or less, but of course not so much as green sand moulds. To know whether a casting has contracted vertically or not, it is necessary to take exact measurements of the mould (not the pattern or sweep), as there is often a difference between the mould and the sweep or pattern.

After the casting comes out, compare the measuring rod or stick with it, carefully note and allow for evident straining. I think it will be found that the casting has contracted as much vertically as it would horizontally, were it possible to have cast it so. There are often castings poured horizontally that, if measured, would not only show no contraction, but would be larger than the pattern they were moulded from, especially if the castings were heavy.

Cores will sometimes greatly prevent the free contraction of a casting. Sometimes light proportioned castings, having cores surrounded with metal, will crack, from there not being body enough of iron to press the cores together. When iron is run all around cores, and the thickness of iron is over one inch, the cores will expand so as to often force iron up through the feeding heads, risers, and pouring gates. Should the tops of the gates get frozen soon after the mould is poured, so as not to allow the expanded iron to come out through the gates, the iron will press against the sides and surfaces of the mould, forcing them outward, and when the castings come out, they are sometimes found to be larger, or as large, as the pattern ; thus showing, as far as measurement is concerned, that there has been no contraction.

The opener or softer the grades of iron, the less contraction there is. Heavy bodies of iron contract less than light ones. The more contraction in iron the more liable are the castings to crack. Castings having light and heavy parts combined or connected always have a strain on them ; in fact, there are very few castings made but have more or less of a strain upon them.

Pattern makers usually allow the same shrinkage on all castings. If they would make a small piece the same size as some part of a mould to be poured, and have the piece moulded and poured from the same iron and at the same time the main casting is poured, they would find that generally the iron contracts differently in the two cases.

The moulder, when drawing these test pieces, must be very careful not to rap them end-ways.

It is not the pattern makers that are to blame for the prevailing ignorance of the different contraction of iron in light and heavy castings. It is the foremen and proprietors of the foundry, for allowing the pattern maker to use the same shrink rule for every pattern he makes.

When a casting comes out too large, the first thing that is thought of is to swear that the pattern maker has made the pattern too large. When it is measured and found to be right, they come to the conclusion that the moulder has let his mould strain too much, which is an admirable corner to crawl out of.

If proprietors of foundries would order their foremen, also their pattern makers, to take measurements for one month of all the different forms of castings that they may be called upon to cast, at least all that are large and of different proportions; also, during this month's experiments, make test bars of all the different grades of iron that are used in the foundry, and after making note of their contraction then try their tensile or breaking strain, something would be learned. To do all this requires no great labor or time. Moulders that take an interest in their trade could, whenever they make a casting of any note, test the contraction and keep a record of experiments that would be valuable.

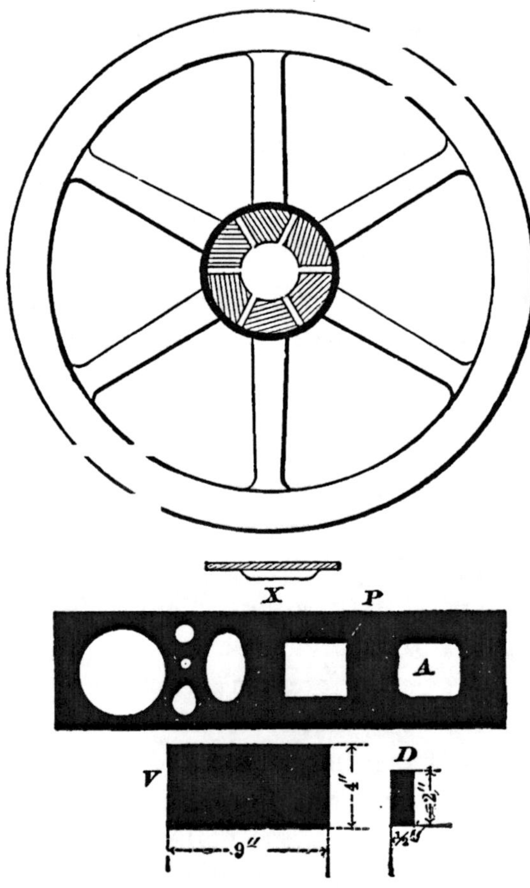

The moulder or pattern maker is not always responsible for the breaking or cracking of a casting. The designer or draughtsman often designs forms that the best of iron and management would not prevent from cracking. Sometimes castings will stand in the shop for weeks, and even months, and, to the surprise of all, will then crack.

Not long since I had occasion to cast a bar 4″ × 9″, and 14 feet long. With the same ladle of iron I cast a test bar exactly the same length, and $\frac{1}{2}'' \times 2''$, both of which are shown at *V* and *D*. The contraction of the thick bar was only $\frac{7}{8}''$, while that of the thin one was $1\frac{3}{4}''$, or a difference of $\frac{7}{8}''$. This gives some idea of the strain that is always in castings that are not made of proper proportions, and also shows the difference there is in the contraction of thin and thick bodies of iron. These simple tests are such as any one can make.

To test the difference of strength between a casting poured with hot iron and one poured with dull iron, I made two bars, and poured one with the iron hot, and then stirring and mixing up the rest of the iron with a rod (so as to work up the impurities in the iron to the surface), until it was as dull as would run with safety, I poured the second one. The next day I took the bars, and resting $\frac{3}{4}''$ of each end on a good, solid iron bearing — the one that was poured with hot iron first—commenced by putting on 50 pounds scale weights in the middle of the bar. Eight 50 pounds were piled on, but before the weight of the ninth one was all on the bar broke.

Now, taking the bar that was poured with dull iron, the whole nine weights were piled on, and an additional 60 pounds' weight, the whole resting on the bar about eight seconds, when it broke. That is, the one poured dull broke with 510 pounds, while the one poured with hot iron broke with a weight of between 400 and 450 pounds. The size of these bars was 1'' square and 4 feet 6'' long.*

Often a man will go to a foundry to get a bid on some plain plates, and thinking that the foundry man will put poor iron into them, he will ask for a low figure. He gets his low figures ; also gets the poor iron in his plates. Common plate castings require as good iron as the common run of machinery castings ; in fact, there are very few castings

* By further experiments on this subject, reverse results were obtained. See p. 8, Vol. II.

made but should have fairly good iron put into them, with the exception of such pieces as sash weights.

Often plates are made with holes to lighten them, or for some special purpose. When holes are made in castings for lightening them, the castings will be stronger for having the holes round, oblong, or oval, instead of having them square. Sometimes, when square holes are made in castings, they will crack, as shown at *P;* whereas, if the corner is rounded, as shown at *A*, it will not crack. A heavy rib *X*, cast on to strengthen square holes, will sometimes do more harm than good, as it causes a strain by making the casting disproportionate.

I have seen castings having on them heavy projections, flanges, or ribs crack, that, were they cast without these parts, would stand a great blow or weight upon them before cracking or breaking. Cracks generally start from some thin or sharp corner, and, when once started, run through the entire body of the casting, the thick portions offering no more resistance, apparently, than the thin. A moulder, when making a new casting, should study the points or sharp corners that will be subject to strains, or disproportionate contraction; and, if possible, have the sharp corners made rounding, and the thin or thick portions made heavier or lighter. Should he be told that it will not do to change these parts, he then, should the casting crack (providing that he has done all that he could to uniformly cool it), is not to be held responsible. Often castings crack that would not, had the heavy portions been exposed to the air as soon as they would admit of it.

Some may say, with better iron the castings would not crack. This is all true enough in many cases. A good, strong iron, having very little contraction about it, would almost make a surety of making a casting of extreme proportions without liability of cracking; but this is not the kind of iron we usually have to deal with. Of course the pig-iron merchant tells us his iron is possessed of all these qualities, but the founder is the one best fitted to decide whether it is the best iron, for by making an analysis or re-melting some of the iron in a small cupola he can readily determine its physical qualities. The first thing a founder should do to discover the cause of a casting cracking is to look at the method of moulding and cooling and not always blame the iron, since to this cause is due the cracking of many castings far more than the quality of the iron.

The cut shown of a fly-wheel, having the hub split between every arm, is a good plan to adopt when making large pulleys, fly-wheels, or gears having cast-iron arms. A wheel made in this way can be relied on. Each arm being a separate casting by itself, when it contracts it is free from any strain or pull, as is the case when the hub is made in one solid casting. Castings made this way are banded with wrought rings, and the opening in the hub filled with babbitt.

When a wheel of any description has cast-iron arms, rim, and hub all one piece, there is generally a strain on the arms or rim of the wheel, in extent depending on the proportion of the rim and arms. If the rim is light and the arms heavy, we may look for a cracked rim, caused by the thin rim contracting faster and more than the arms. Again, the arms will be the part to crack, caused by the rim being too heavy. In either instance this generally happens while the casting is yet hot and in the sand.

In the case of pulleys, etc., having the arms crack after the casting is taken out of the sand, we have a more complicated state of affairs to deal with. In pulleys having heavy arms, compared with the rim, we often see the heavy arms cracked, the light rim remaining whole. In looking at such castings it will be observed that the hub is heavy in proportion to the arms and rim. This hub is the last portion of the pulley to become solid; the rim, being light, has become solid, and is already contracting, driving before it the half-molten arms into the yet liquid iron in the heavy hub. When the hub solidifies, it contracts, pulling with it the arms, causing a strain, which, when the pulley gets a slight jar, will make the arms crack at the weakest point.

The same principle is involved in light-armed pulleys as in heavy ones; that is, so far as the heavy hub is concerned.

When there is a heavy hub required, it should be cooled as soon as possible by stripping around it, taking out the core, and cooling with water.

Above everything, as regards the contraction and cracking of castings, we should not forget that a thin body of iron will contract more than a thick body, and, whenever there is a casting formed disproportionately, there is always more or less strain on some portion of it. And also massive castings are subject to exterior and interior strains, as will be seen by the following discussion·

The question is asked, Why a heavy body of iron will not contract as much as a light one?

Knowing such to be the case, it must also be acknowledged that there must be a cause, and as this is one of those subjects that practical tests can very seldom be applied to, the following theory is presented:

When castings cool, some of their parts always cool faster than others, and the parts that cool the first are the exterior or outward portions. Often there are castings cooled solid on the outside, while the inside portion is perhaps in a molten state. To discuss this question we will take the size of the castings shown in the cut, one being 4″ × 9″, and the other ½″ × 2″, the moulds for each being 14 feet long. These were cast with the same iron, and at the same time. Now let us watch the process of cooling.

The light one soon commences to cool, and we see it contracting. The outside portion of the thick casting commences to cool, and endeavors to contract also, but it cannot. We look at it to determine the reason, and inside this cooling crust we know that it is very hot, and the further towards the center we go the hotter we find it. The inner portions of this casting we know are not yet in a state to contract as fast as the outer portions, and when a casting becomes entirely cool its contraction ceases. In this casting some parts may become cool, and still all parts not have contracted as much as the nature of its iron requires. There is a certain amount of tensile qualities about iron that permits its molecules or particles to be stretched to a certain limit, and when this limit is exceeded the result is a cracked casting.

Returning back to the cooling casting, we find that the slower interior cooling iron will not allow the faster exterior cooling iron to contract as much as it should, according to the degree of heat it has lost, and by this cooling process we have the exterior portion of the casting contracted by forces which hold it back from contracting as much as it should. Now, when the interior portion contracts, it finds the same resisting forces to prevent its natural contraction, the exterior having lost most of its heat, and therefore having contracted about all it can, will not permit the interior to contract any more than the exterior; and thus, as one holds back, so does the other, and the result is that the $\frac{1}{2}'' \times 2''$ casting, not having these conflicting forces to contend with, contracts about all that the grade of iron composing it naturally calls for, while with the thick casting, $4'' \times 9''$, we find the contraction just about one-half of what it should be if it had been as free to contract as the light casting.* Often when looking at solid massive cast-

* Another force at work is that lying in the fact that slow cooling causes iron to be open-grained, while fast cooling is the reverse.

ings, cracks are seen running from 3" up to 8" long, and about ½" deep; these cracks are usually in sharp angles, and their origin can seemingly be attributed to the strain there is upon the exterior portion of the casting, caused by the law that apparently governs the cooling of thick bodies.

Iron, when changing from a liquid to a solid state, is said to become a mass of crystals, which assume different forms and sizes, being regulated by the length of time the casting takes to cool, and the temperature of the iron when poured into the mould. The lines of crystallization are very seldom visible to the eye, except in the cases of chilled iron. In castings that are not chilled, the lines of crystallization depend upon the direction in which the heat passes off the fastest, and with the least resistance. In regard to this formation of crystals, MALLET observes: "It is a law of the molecular aggregation of crystalline solids, that when their particles consolidate under the influence of heat in motion, their crystals arrange and group themselves with their principal axes, in lines perpendicular to the cooling or heating surfaces of the solid."

Sharp angles, corners, projections, and squares, combined in castings, cause them to show lines of crystallization run in different directions from some given point, and this point, from which the lines of crystallization connects, appears like a rope with a number of small strings tied on it, and pulled by unseen forces in two opposite directions. This rope or section of the casting, from which all these strings or lines of crystals radiate, is the weak point of a casting, and there are very few castings so shaped but that many such weak points appear in them, but for the shape no one can be fairly blamed. The moulder, even if he thoroughly understands the problem of crystallization of different-shaped castings, could very seldom in practice cool a casting so as to cause the crystals to radiate in lines, strengthcuing the weak points, or change them from their natural course. If there is to be responsibility placed on any one, the designer of castings is the one who should assume the greater portion of it. But since he cannot always have a design made so that the shape of a casting will allow the lines of crystallization to radiate in a manner which will least weaken a casting, we should be very careful how we censure him. The contraction and cracking of castings is a subject that has many perplexing features; and it is hoped that this chapter will prove of assistance to all those who are in any way interested in the production of castings.

FEEDING AND SHRINKAGE OF MELTED IRON.

THE assertion often made by writers, "that melted cast iron expands at the moment of solidification, so as to copy exactly every line of the mould into which it is poured," always sounded to me very odd. The word "moment" would imply a sudden dividing line between liquid and solid iron. In melting iron there is nothing sudden from the time it leaves the melter's hands till it is tapped out into a ladle, from which time it cools gradually, the time it will take to cool depending on the shape and size of the mould that the iron is poured into. The amount of expansion or shrinkage will be according to the grade of softness or hardness of the iron, and the quality of the ores that the iron is made from.

Soft iron is open grained, and when melted has more life than hard iron, and may expand some in cooling. Hard iron is close grained, melts quicker, and has more shrinkage than soft iron. Melted iron, when cooling, cools the fastest at the bottom of the mould and at the sides and cope surfaces, which draws molten iron from the hottest or central portion to supply the shrinkage of the cooling parts. If this central, or last portion of the iron that cools, is not reached with a feeding rod and hot iron to supply the shrinkage, the last parts to cool will be honeycombed or hollow.

There are often castings that cause explosions or breaks through the thickest, and, as is thought, the strongest parts. One cause for this is that the castings were not fed in the right manner. There are often castings that would require a half dozen feeding heads to make all the parts solid, and in some cases the designers of machinery or the pattern makers will have castings disproportional, so that the thick portions cannot be fed, or the shrinkage of the heavy parts supplied with iron. Such parts would be stronger if they were made lighter, and had what iron there was solid, rather than heavy, with a honeycombed center.*

The cut *X* shows a round die block that was used in a lamp manufactory for pressing a composition of metal. There had been several of them made, but, being unsound, the pattern was taken to another shop. When the pattern was shown to me and the trouble explained, the first question I asked was if there had been a large feeding head used, and if they were fed well? I was answered "Yes." Further questioning showed that there were no signs of dirt or holes until the casting had from ½" to 1" turned off it. I gave the job to a man I knew was not afraid of a hot job, and saw that he got hot iron when wanted. He stuck to it until his rod was driven up into the feeding head by solid frozen iron below it, and I had the pleasure of telling him that his casting was the best and solidest the machinist had ever finished up. What the machinist called dirt, or holes in the bad castings, was only honeycombed or porous iron caused by improper feeding. One moulder will feed a heavy piece of casting in half the time another will take, and still, to all outside appearances, have it solid; but should the casting be cut up into small pieces, it would appear that he did not feed his casting *solid*, as the iron in the part that remained hot the longest would be liable to show holes or be very porous. Take, for example, solid castings one foot in diameter, and let them be cut through the middle, or the outside all be turned off until they become balls 3" in diameter; it would be safe to say that they would present a very rotten appear-

* A valuable chapter to be read in connection with this is found on p. 1, Vol. II.

ance, that is, if they were fed as solid heavy castings are generally. The length of time that it will take for some liquid castings to become solid throughout is very often longer than is supposed, and in many cases moulders should modify their assertions of a casting being fed solid.

In setting a feeding head on most patterns it should, if possible, be set on the thickest portion of the casting, or that part of it which will keep the longest hot. The feeding head should be of such size that it can be kept open, with hot iron until the casting is set. In starting to feed a casting, the rod should be put in slow and easy, and if the mould is not too deep, it should touch the bottom, and then be raised up two or three inches, so that it will not be punching holes in the mould. Some moulders, when feeding, work their rod up and down in the center, and the sides freeze up and close or solidify while the iron in the mould is yet in a liquid or molten state ; or they will put a small feeder on, that will freeze so quickly that they cannot get a rod into it, or if they do, it will stick fast, and then they will complain of some one for not bringing hot iron when wanted. A moulder should seldom make this excuse. He should have the feeding rod hot, and the head the right size ; and instead of working the rod in the center, work it up and down around the sides, so that the freezing iron will be pushed or worked down into the casting, and the hotter iron in the casting worked up into the feeding head. This keeps the head and casting at the same temperature.

When you do get hot iron, always have a hole worked in the head to hold as much as possible, so that it will help to cut away the freezing iron on the side of the head, making the iron in the head hotter than in the casting. Put in hot iron when there is a good chance to get it, and don't call for it just as you see the cupola man going to stop the cupola up, or do some other as sensible trick, as putting in the rod

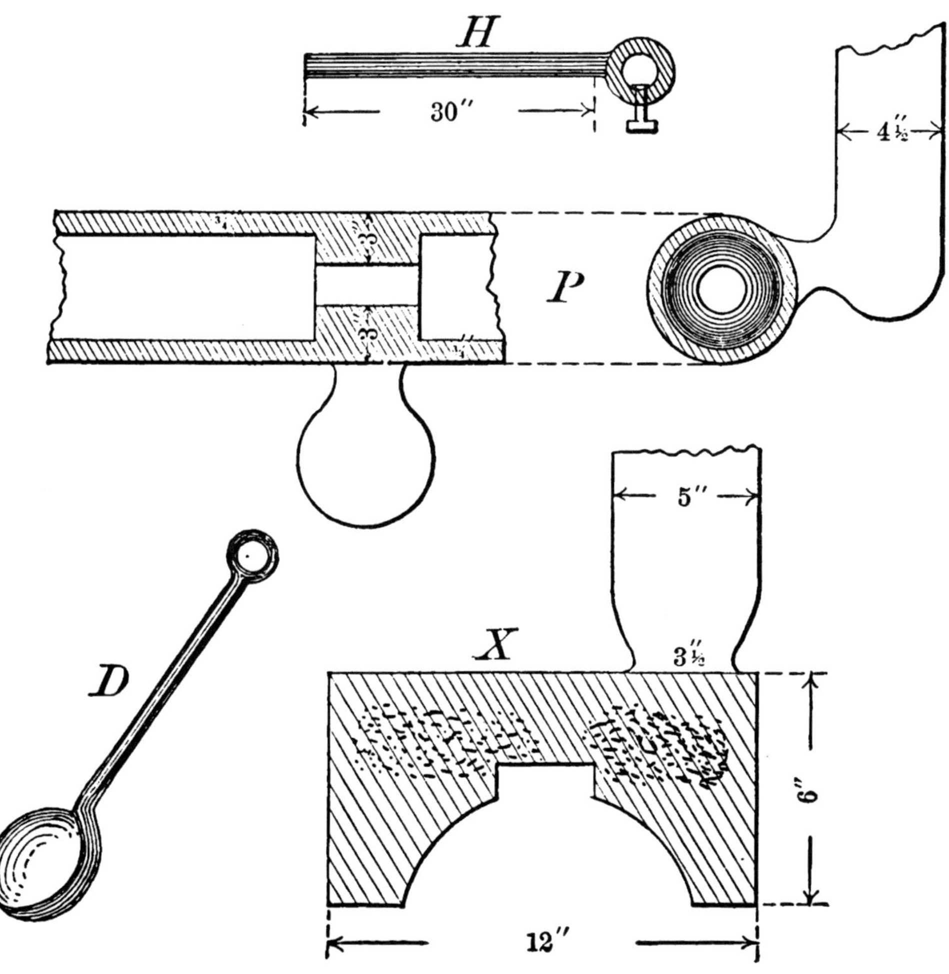
H
30″
4½″
P
3″
3″
5″
D
X
3½″
6″
12″

so that it does not go into the casting, which lets the neck of the feeder freeze. The rod should be kept down into the casting, and let the iron as it freezes at the bottom push it up out of the casting.

There are often cases where the iron or wooden bars of a flask will not admit of a proper-sized feeder. In such cases where the bars cannot be readily widened, the feeder should be built up in length, to make up for the loss in diameter. A small feeder closes quicker and takes more hot iron to keep it open than a large one. Large feeding heads require less work and attention than small ones. With feeders 10″ and upwards, as soon as the casting is poured, put in the feeding rod and work it around for a minute or two to work the dirt up to the surface. Then take the iron dipper, as shown at *D*, of which a shop should have three or four sizes—the dished part being about 2″ deeper and the handles about 4 feet long—and dip out the dirt and as much of the dull iron as is necessary to make room for one or two hundred pounds of hot iron. Then work the rod to mix the hot and dull iron, and throw on some blacking to keep in the heat. After fastening on the holder *H*, or a pair of blacksmith's tongs, to hold up the rod, the moulder can rest from ten to thirty minutes, occasionally lifting up the rod to see how the iron is, and that it is not freezing at the smallest part or neck of the feeder. As soon as the neck shows signs of closing up work the rod around to open it and to mix the iron, after which dip out some of the dull iron and pour in some hot iron, and cover again with blacking. Repeat this operation until the iron in the casting commences to stick to the rod, then the moulder should give it all his attention, as the neck will close up if the feeding rod is not kept in constant motion, especially if the iron is hard.

I was at one time foreman of a shop where rolling-mill work was done. The proprietor being a moulder, and knowing the failings of some moulders, adopted a plan of feeding his large and small rolls a certain length of time. Rolls weighing about four tons he would only allow to be fed 70 minutes. It made no difference whether they were poured hot or dull. When the time was up the heads would be filled with hot iron and the rods worked to open the neck, then taken out and the iron covered with blacking. The plan was a good one to accomplish the desired end.

When a proprietor or foreman thinks that he has no men who understand feeding, or who will stick to a hot job, the best and surest plan would be to make the feeding heads without a neck; that is, having the feeder the same size at the bottom as at the top. Then let the head be cut off in the lathe, as is done in the manufacture of cannon or large guns. In such work the gun is cast from 2 to 5 feet longer than wanted, the extra length answering for a feeding head.

In writing this article, the subjects chosen are castings that cause trouble from shrinkage, and are good ones to show the principle of feeding and shrinkage.

The cut *P* shows a broken pump of the kind used on a locomotive, and when the castings were bored out they would be porous and dirty in the heavy section. To remedy this they were cast in dry sand, and on end, but with no better result. The thickness of iron on each side of the heavy part is only $\frac{3}{4}''$, and the heavy part 3″ thick. My attention was called to the job, and seeing the trouble, I made two or three castings in green sand, and with a feeder cut into the heavy section, as shown. The casting bored out solid and clean. Casting them in dry sand, and on end, would not make the thick part sound, as the thin part would freeze before the thick, and the iron to supply the shrinkage would have to be withdrawn from the upper and hottest portion to supply the shrinkage below; so that in boring the heavy section the part that was down would be solid, and the upper part porous or honeycombed.

I have often seen castings go out of the foundry that were required to stand heavy strains or pressures, which, if the party that received them had understood the feeding of melted iron and had seen them fed, would never have been accepted. It is not altogether ignorance on the part of the moulder, but the desire to get rid of a hard and hot job, that is the main cause of ill-fed castings. There are very few heavy castings that are fed so as not to be somewhat porous. In such castings as levers, etc., that must stand strains or blows, the point or section of the castings which has to stand the greatest strain is not generally the point that feeders should be put at; it is better, in case of a long lever, for instance, to set the feeding head at one end, if the end or portion is thick enough to be kept in a liquid or molten state as long as the heavier parts that receive no hot metal. Where feeders are set is generally a point of weakness, as this point is not the same grade of iron; and it is also from this point, on account of its being supplied with hot feeding iron, that the other portions of the casting draw iron to supply shrinkage; and if the greatest of care and judgment is not used there is always more or less danger of there being a porousness or holes below feeding heads, thus causing that part of a casting to be weak. Even when the casting has been fed perfectly solid and compact, the mixture of a foreign grade of iron at this point is sufficient to cause a weakness. When setting feeders on a pattern the moulder should know whether the casting is required to have a solid finish, or if strength is required. The close reader and thinker will see that there is a difference.

BURNING OR MENDING HEAVY CASTINGS.

THE principle employed in the process herein described for burning a new neck and wabbler upon the end of a broken cast-iron roll, such as is used for rolling iron, steel, and other ductile metals, may, with a display of moderate skill and judgment, be practically applied for burning or mending a variety of heavy broken castings.

The object of this process is in mending heavy castings, to avoid the expense of making new ones, and, if properly performed, it is very economical, and will save much time, labor, and expense.

The most essential points to be observed in mending heavy castings by this process are as follows: The melted iron must be very hot, and of a medium soft quality, for the hard iron chills quickly, and therefore does not perfectly cement or unite with the broken surface of the casting.

An outlet must be provided to allow the melted iron to escape from the mould as soon as it is poured in, partienlarly at first, as the hot metal is liable to chill. The hot metal should fall directly upon the surface of the fracture, and, after beginning to pour, a uniform, steady, cutting stream of iron should be kept flowing from an elevation as high as possible. In burning the neck and wabbler upon a roll, the larger the roll the more successful will be the operation, on account of the larger surface for the melted iron to burn or cut into, thus uniting more perfectly.

As a preliminary operation, the roll might be made as hot as possible in the foundry oven, after which it should be lowered into a hole previously dug in the foundry floor, keeping the broken end even with the level of the floor, as shown in the engraving upon page 269, after which the hole should be quickly filled with sand, rammed up solid, particularly around the top of the roll. The surface of the fracture should be chipped all over, so as to break the skin of the metal and remove the rust.

The wabbler and neck must be moulded in dry sand, for if made in green sand the falling melted iron would cut the moulds all to pieces. These moulds should be made in sections, as shown in the illustration, where 2 represents the mould for the neck, and 3 the mould for the wabbler, while 4 is for the riser or feeding head.

The flask 2 has three places cut out at the bottom edge, *A*, *B*, *C*, which are for openings to allow the melted iron to freely escape while the neck is being burnt on. As soon as the roll has been securely placed, the flask 2 should be put in position, care being observed to keep the opening for the neck exactly in the center of the roll, to allow for turning up the journal in a lathe. Pigs of iron or other suitable heavy weights should be placed upon the handles *X*, *X*, to hold down the flask. If an air furnace is used in the foundry, three basins should be formed, one opposite each of the openings *A*, *B*, *C*, to catch the waste iron; but if there is no air furnace at hand, then a pig bed should be made, throwing the spare sand around the joint between the roll and flask, so that there can be no run-out. The sand should be cleaned away from the outlets, and runners made leading to the places formed for catching the waste iron. Sometimes, if there is a mould that can stand pouring with dull iron, a ladle may be sunk down into the floor to catch and save the iron. If the iron should be likely to run upon the face of the roll at the outlets, it should be smeared with oil at those points to prevent the melted metal from adher-

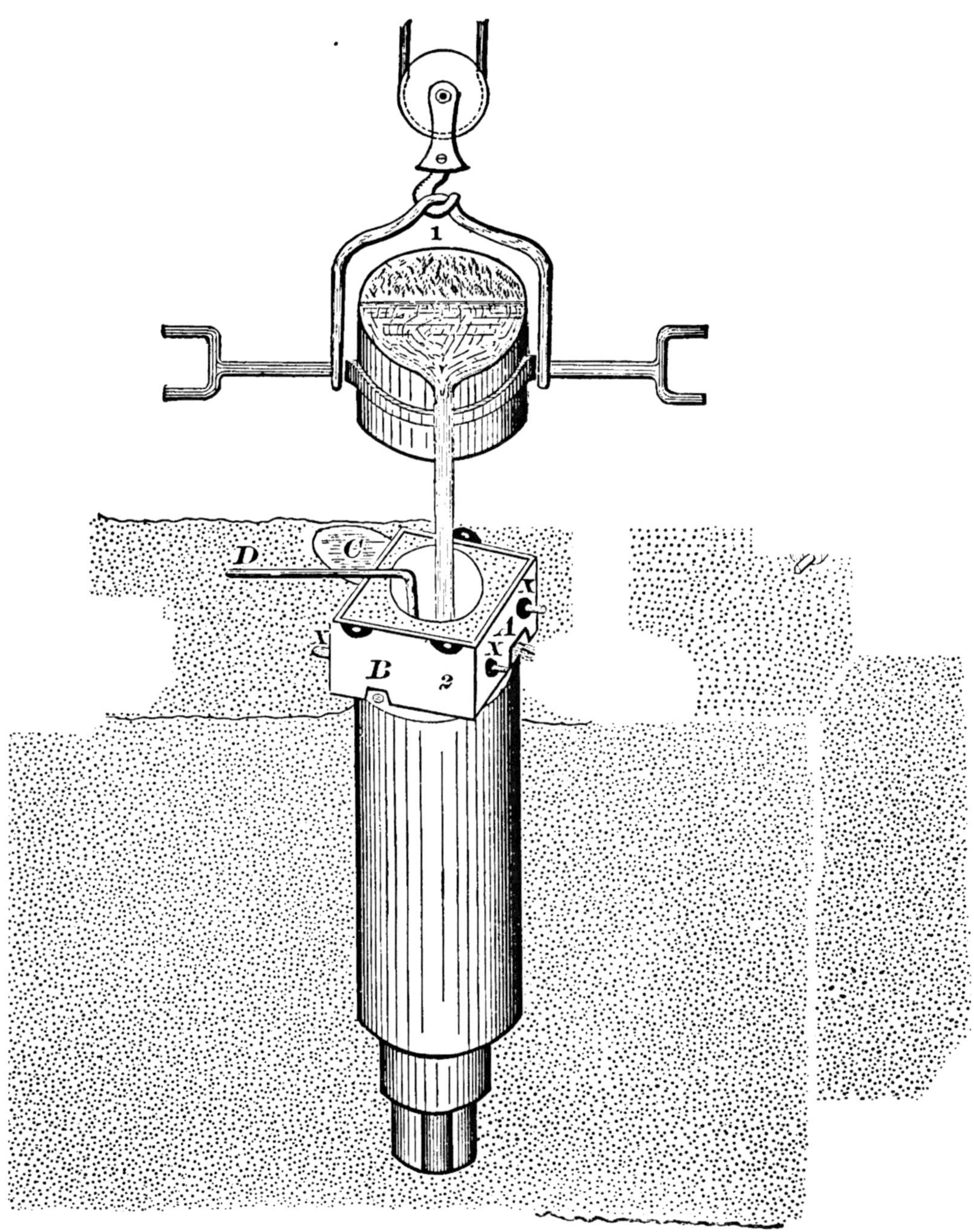

PROCESS OF MENDING HEAVY CASTINGS.

ing to it. In commencing to pour the iron, the ladle should be held very low, and then gradually raising the ladle until the metal will have a fall of about four feet. The ladle is supposed to be handled by a requisite number of men not shown. The man in charge of the process can easily ascertain if there are any places upon the surface of the fracture where the melted metal is not cutting, by means of the bent rod *D*, and have the molten stream directed upon those places. Some small pieces of tin or zinc should be kept at hand, and by constantly throwing them into the mould the iron in the holes that are burnt is thereby made hotter than it would be otherwise.

The iron should be poured until only about five hundred pounds remain in the ladle; the openings, *A*, *B*, *C*, should then be stopped with a stopping stick and clay; then fill the neck neatly full of metal.

A protecting ring, not shown (which is a thin plate of metal, having a hole in the center, the same size as the neck, and is placed on top of the flask to prevent the falling stream of hot iron from cutting away the edge of the mould), should be lifted off at once by means of handles attached to it; then skim off all the dirt and slag from the surface of the metal, after which the wabbler mould 3 should be quickly placed upon the flask 2, and the feeding head 4 upon the top of the flask 3. These flasks should be securely held by placing heavy weights upon the top, after which the mould should be quickly filled and fed with metal until it solidifies. If the neck has been burned successfully, the old or original neck will generally be the first to break. In regard to the amount of iron necessary to complete the operation described, I may say that it depends entirely upon the nature of the casting to be mended. For a roll weighing from 5,000 to 7,000 pounds, it will take about 1,800 pounds for burning on wabblers or neck, while on small rolls, about 1,200 pounds of metal will be required. The cost of the process is comparatively small, and saves the trouble and expense of turning up a new roll, besides effecting a great saving in time, which is a very essential point in rolling-mills.*

---

* For further valuable information upon this subject of "burning," see chapter on "Welding Steel to Cast Iron, and Mending Cracked Castings," p. 217, Vol. II.

CHILLED CAST-IRON CASTINGS.

The surface part of a casting that is wanted to retain a certain shape, size, and smoothness, and to withstand a constant wear and tear, can, in most cases, be chilled when cast by having iron to form the shape instead of sand. The iron mould or chill, when made of cast iron, should be of the best strong iron, having very little contraction, as the sudden heating of the surface by the melted iron is liable to crack it, or in a short time the face will be full of small cracks or raised blisters. When melted gray iron is poured around or against the surface of solid iron, it is chilled from ⅛″ to 1″ in depth, depending on the hardness and closeness of the iron the mould is poured with. In order to chill this iron as deep as 1½″ and upward, there must be some cast steel or white iron melted with it in the cupola. The proportion will depend on the quality of the iron and steel used. Steel borings can be put into the ladles and let the hot iron mix with them; but the best plan is to have some old steel castings, or pieces of steel rails, and melt them in the cupola, and when the iron is in the ladle, mix or stir the metal with a large rod.* With strong, close iron, about one part of steel to five parts of iron will often work well. *Iron for making chilled castings should be strong, as chilling iron impairs its strength.* The founder guided by the analyses of pig, instead of its fracture, in selecting it, will be most successful.

I had reason at one time for studying the cause of chilled castings being bad. I was working in a shop where they made some small chilled rolls, about 10″ in diameter and

---

* Valuable information upon melting and mixing steel or wrought iron with cast iron to obtain strong or chilled castings is given on p. 217, Vol. II, and in "Metallurgy of Cast Iron."

14" long. The thickness of chill for chilling the roll was 3½". The top and bottom necks and wabblers were moulded in dry sand. The job was given to me, and I moulded and cast three at a heat. Out of the first three there was only one good one, and I was told that I was lucky at that, and that the proprietors would give a good deal if they could make their own rolls, as they had to send away and pay a heavy price for them.

These rolls had to be chilled 1¼" deep, as there were grooves turned in them for rolling bar iron, and when the grooves got worn out of true they were turned again. After studying the trouble over, I cast three more with good results, and from that on there was no more trouble with them.

Melted iron, when poured inside of a chill, similar to a roll or car-wheel chill, cools and forms a shell in a very short time, the thickness of which will depend on the hardness and temperature of the iron. In small rolls and wheels it is during the course of the first two or three minutes that the checking or cracking generally takes place; for as soon as melted iron commences to freeze, it starts to contract more or less, and as the shell thus formed becomes cool, or solidified, it contracts, so that the contracting shell has to stand, or hold in the pressure of the liquid iron inside. Should the mould not be dead level, the inside liquid metal will have the most pressure at the lowest point of the shell, and often cause this part to burst open. A check or crack never starts at the top part of a mould, but always at the bottom, and if you look closely at one of these cracks you will see it is the largest at the bottom and running up to nothing. In some cases you can see where the inside liquid iron has flowed out, and partly filled up the crack.

I have often asked car-wheel moulders why it was they cooled their iron to a certain temperature, and they would answer: Because it keeps the wheels from cracking. Some days, when they would have three or four wheels cracked, when asked what was the matter, they would say that the melter did not mix his iron right. So far as mixing the iron is concerned, it will stand a deal of excuses; but it is not always right to put the blame on the iron for three or four bad wheels out of a heat of sixteen. If the moulder would make a straight-edge that would reach across the top and come down on to the turned level face of the chill, and then level his flasks, instead of dumping them in any shape, the poor melter would not get blamed so much as he does for cracked wheels.

In making chilled rolls the temperature of the iron is as important a point as it is in the manufacture of car-wheels. It should be of fair temperature. The duller the iron the quicker is the outside shell formed, thereby offering a stronger resistance to the pressure of the inside liquid iron. Of course, the moulder must use his judgment in cooling off the iron, for if too dull the face of the chilled part will be cold shut, and look dirty. The rolls should be poured quickly at the bottom neck and the gates cut, so as to whirl the iron and keep all dirt in the center and away from the face of the chill.*

When the mould is full and the iron seems to require feeding, from the way the feeding head settles, do not put in the feeding rod until the neck is about to freeze up. When you do put it in, don't ram it down suddenly, so as to cause a pressure on the contracting shell, which would be liable to crack it. When feeding, work the rod slowly. In Pittsburgh, a great center for the manufacture of chilled rolls, they do not feed their rolls at all. The feeding heads are made long, without a neck on them; they are made the full size of the wabblers, and then cut off in the lathe. In pouring the rolls the iron is taken from an air furnace into a large ladle, and, after being cooled to the required temperature, the iron is poured into the moulds by

* For a good plan to gate chill rolls, see p. 238, Vol. II.

basins that are made very large, so as to be able to keep the dirt out and have the iron go in fast. As soon as the mould is filled to the top of the wabbler the pouring is stopped, and then the balance of the feeding head is filled up with hot iron, carried in hand-ladles from the cupola. Some of the large-sized rolls have feeder heads from three up to four feet long cast on them. The heads are made nearly the size of the body of the roll, from 3″ or 4″ above the top of the wabbler up to the top of the head. As soon as the heads are filled up with iron, they can then be covered with sand, and they will then feed the casting without any further handling.

It is better to make the chills hot by heating them in the oven, the iron will lay closer and make a smoother casting against a hot chill than when poured against a cold one.

By having the mould dead level the pressure will be equal all around. Whenever there is a check or a crack, it is assisted by unequal pressure of the confined liquid metal against the contracting shell; and, whether some moulders believe it beneficial or not, to have a chill mould level, when being cast, they can but, upon studying the elements here advanced, acknowledge that when a chilled casting checks or cracks, the question of greatest point of pressure can have much to do with it. Although castings are well cast with moulds not level, the worst cases would be less if the moulds were leveled; and many chilled castings that are lost through checks or cracks would be often saved. Checks or cracks may also be often caused by reason of the contraction pulling the cooling shell away from one side of the chill before the other, thereby giving the inner fluid metal a chance to exert its pressure at the side first leaving the chill. Factors causing this effect may often be found in "fins" being held between the joints of the mould, or gates, holding the casting more rigidly to one side of the chill than the other. Further information on this subject, and plans for making chill roll flasks, will be found on p. 234, Vol. II.

MAKING CHILLED CASTINGS SMOOTH.

To make chilled castings without having them streaked or cold shut is a very important feature, and one that has caused a deal of trouble in certain classes of work, such as anvil or die blocks, which are usually cast with the chill lying horizontally. When the moulder pours his mould he will start slow and easy, being afraid of spilling the iron or cutting the runner; and, when the casting comes out, it will not have a smooth face. The excuse will be dull iron, or the men did not stand still, or did not hoist the crane when told to do so, or something else. "He is a poor moulder who cannot make a good excuse," is thought to be a good adage among moulders, and I think it is, too; for there are some moulders who would not sleep two nights in the week if they thought their excuses were discredited. *In point of fact, it is more consoling to the mind to frame an excuse than it is to study the cause and find the fault is in our own ignorance.*

Whenever a moulder has trouble with his work, he should study the cause before making the piece the second time. Any moulder who follows this plan knows its value, not only in making good work, but in enabling him to understand *cause* and *effect*, and the principles of his trade.

In studying a plan for making flat chilled faces smooth, I made the runner and gates large; and if a crane ladle were used, the basin should be made large, and the top of the runner have an iron cone plug (as shown at *W*, page 163) to close it up; and, when the basin is about full of iron, lift up this plug, and in goes the iron with a rush, and immediately covers over the face or surface of the chill with a body sufficient to keep down any tendency of the iron to bubble, caused by hot iron coming in contact with cold. When iron is poured so as to run into the mould as it first comes from the ladle, the bottom is covered slowly, and the casting is generally sure to look streaked and dirty on the face.

There are blocks and dies cast flat that could be cast on a slant, or perpendicularly, thus causing the iron as it is poured to cover or rise on the face of the chill in a body. When iron is poured or run so that it immediately covers any section of the chill that it strikes, there is very little danger of the casting being cold shut or streaked, unless it should be caused from using a new chill or one that has not been used for some time, or from the effect of bad oil or too much of it. Some will often take new chills, or those that have been lying idle, and pour some melted iron on or in them to heat them before they are wanted. This will burn off any rust or scale that may have collected on them. When oil is used for rubbing the face of a chill it should be light and clear, and very little should be used. When too much oil, or when black, heavy oil is used, it burns when the hot iron comes in contact with it, forming a heavy gas that will throw the iron back from the face of the chill, and cause it to bubble, and liable to generate a dirty scum that will mix with the iron and make the face of the casting look dirty. Whenever I use oil for this purpose I use coal oil, as it is light, and, I think, the best. The use of oil on the chills is to preserve them and keep the hot iron from cementing or cutting into them. On castings that the iron as it runs into the mould does not strike the face of the chill, you will often get a smoother face by not using any oil at all, provided the chill is in constant use.

The proper temperature of the iron when poured will depend on various circumstances, and the moulder must use his judgment in this respect. This temperature has much to do in many instances with the checking or cracking of castings, so that in pouring them with hot iron, while the prospect of getting a smooth face may be better, the danger of losing the casting from checking will be materially increased.

SPLITTING PULLEYS AND OTHER CASTINGS.

It is often necessary to cast pulleys, gear-wheels, fly-wheels, etc., whole, and split them in halves after casting, for convenience in handling or fastening to a shaft. The splitting of such castings has not always been a success, from the effects of which both moulders and machinists have been annoyed. There are several ways of splitting such castings, and what would be good for one style would not do for another. I have seen wrought-iron plates set in the mould to split heavy and light castings, and although the plates were painted with blacking, coal-tar, oil, or rosin, the iron would eat into them, so that in trying to split the casting it would be broken, or it would take much time and labor to split it. A plan that I found to work successfully for such castings, was to use two plates instead of one, and if it is preferable to have the plates stick to the casting, so that in fitting together it will not be necessary to handle loose plates, it would be better not to paint them at all. If it is thought the iron will not adhere to the plates, holes may be drilled nearly through them, from the sides against which the iron will run ; or the holes may be drilled quite through, if care is taken that they do not come opposite each other in the two plates. This plan will make a sure thing of splitting a casting without any trouble, and when the casting is placed together again, it will generally be the same circle or shape that it was before it was divided, and without any fitting on the part of the machinist. In using these plates, if the casting requires them to be very long, there might be trouble caused by their expansion and contraction ; if they are required to be over two feet long, they could be used by having the plates made in short sections, so as to be set into a mould. For some classes of work cast-iron plates, made the required shape and size, might be preferred, as the iron will adhere to them better, and be less liable to show the joint. There are moulds in which the double plates cannot be made to stand up together. In such cases the two plates may be fastened together with two small rivets near their ends, so that they can be easily opened. The accompanying cut represents the plan and elevation of a split pulley, and shows a good way of using cores for splitting a casting, so that when put together there will be no fitting required. There are very few jobs that the foundry man dislikes more than splitting pulleys from whole patterns. Whether they are made from a draw or a split pattern makes very little difference. Not many years back it was thought a great favor to get a pulley cast in halves. Nowadays they are ordered the same as other castings, and, as a rule, they are expected at the same price as plain pulleys, although it takes about twice as long to make them. I have said to parties, "I can show you some nice split patterns, just the size you want every way," and their answer would be, "O, I guess you had better make them from the whole pattern, and split them, as it will save me time in planing and fitting." Of course the result would be they must be made as wanted, or the custom of the parties lost to the foundry. Having split pulleys in almost every way, I think the plan as herewith shown to be good. The hub of the pulley shows a lug, *X*, *X*, on each side. A core from $\frac{1}{4}''$ to $\frac{3}{4}''$ thick cuts through the top and bottom. The iron at the edges of the core should be of a thickness to crack open easily, and still have bearing enough to hold the joints strong when bolted together to be turned and bored out. There are two styles of lugs, or ears, shown on the rim for bolting the halves together. The round, separate lugs *B*, *B*, are made in a core box, and when ramming up the pulley, set the dry sand lug core against the pattern, and ram them up. The opposite lug is made with a piece of pattern with core print on the bottom, to set in the long, flat core, as shown at *D*. This splitting core *D* is sometimes run half-way into the rim of the pulley, in the form of a sharp V; this will make the splitting of the rim easier. This latter form of lug is used for light and heavy pulleys. In splitting these pulleys the job is generally left to the machinist, and I think the best plan is to split them open before boring or turning, as the chisel marks can then be all turned out, and a truer pulley made than by splitting after finishing.

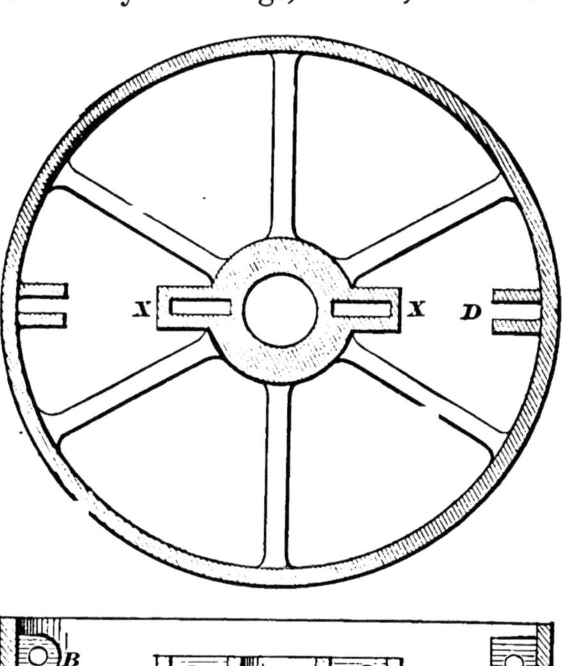

STRAIGHTENING CROOKED CASTINGS.

'WHICH way will my casting go?" is a question that is often asked. To answer this question correctly every time would be a very hard task. The moulder may be very well versed in the cooling and contraction of castings, but sometimes a complicated piece will come along that will puzzle him. In some castings the sooner some of the parts are cooled the less liability there is of the casting being crooked, while again there are others that require to be cooled from some certain temperature.

There are two reasons for castings warping. The first is ill-proportioned thickness; the second, the allowing of some parts to cool before others. To these a third reason might be added, viz., the quality of the iron. Bad proportion is, perhaps, the chief cause for warped castings.

The two sketches illustrate the straightening of a crooked casting, and also how gates and runners often make castings crooked.

Castings for house work give trouble from warping, as they are generally light and long. When the thickness of metal in a column is not equal, the thin side will generally cool first, and draw the thicker side toward it, thereby making the thin side of the column concave and the thick side convex. And the thick side being the last to cool, will often draw it back, thereby leaving the casting, when cold, concave on the thick side. A variation of $\frac{1}{16}''$ in thickness is often enough to make a column crooked. When care is not taken to have the thickness equal and the core well anchored, crooked columns may be looked for. Cores that are made in half-core boxes, and then pasted together, should always be calipered before setting the bottom chaplets, to see if they are round. Should the core be found to be flat or out of round, the difference should be divided between the top and bottom chaplets.

Moulders will sometimes set a flat core on the bottom prints, allowing for no variation, and when the cope is set on there will be, perhaps, more thickness of iron on the cope than on the bottom side. If questioned about this by the foreman, they will answer that the core-maker should have made his core round instead of flat. This is all true enough; but as it is a well-known fact that very few cores are pasted together so as to be exactly round, like a core that is swept up with loam or green sand, the moulder should always

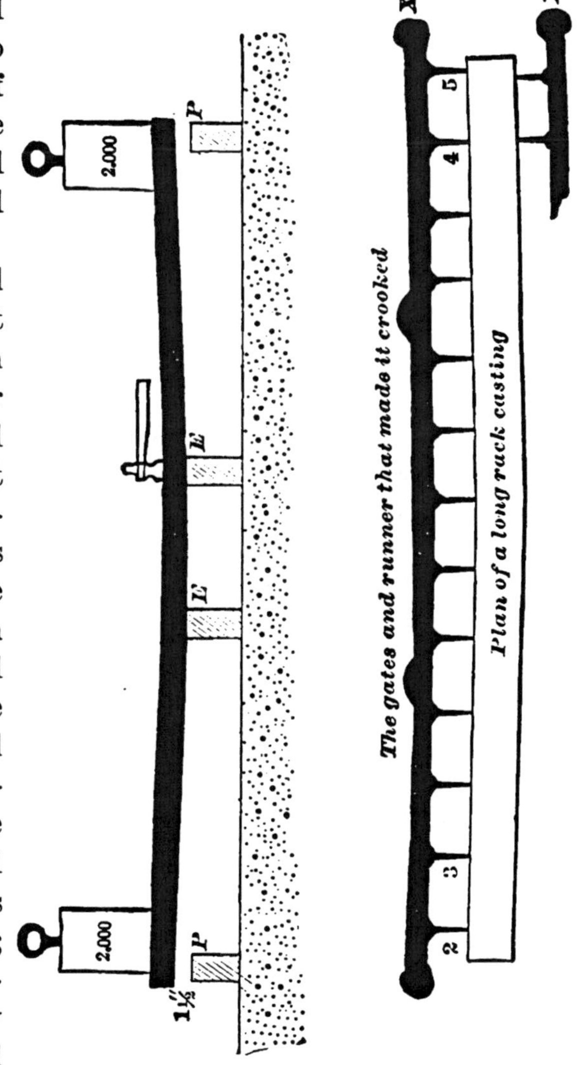

caliper them, and then the pattern, and set the chaplets accordingly.

There are also many house-castings, such as lintels and ornamental work, that are troublesome from having flanges and ribs, for the purpose of strengthening them and saving iron in the main casting. The flanges or ribs are usually either thinner or thicker than the main casting. It can only be told how a flange or rib will draw a casting in the case of a special pattern; that is, no general rule can be given. The same shaped flange or rib attached to different-proportioned castings will not make the castings all draw in the same direction. Although it is mainly flanges and ribs that are the cause of many castings bending into all kind of shapes, we must not lose sight of the fact that the main casting is just as much responsible. Either (the flanges and ribs or the main casting) could generally be cast separately and come straight. In all cases the thick side of castings does not require cooling. Sometimes columns can be made thicker on the cope side, and this thickness be the means of preventing the casting from being crooked, in the bottom the iron is there the closest grained and soundest, while the upper surface is more porous and contains most all the dirt, which allows the heat to escape much faster, the cope covering of sand offering but little resistance to it. The cope surface of a casting generally cools the fastest, especially when the copes are shallow. Before endeavoring to cool a casting, we should consider the greater amount of heat lost by the *upper portion than lost by the bottom part.*

About all that can be said as regards the cooling of ill-proportioned castings, to keep them straight, is: *cool the thick part and keep the thin part hot, so as to have the same temperature in each portion.* Keeping all parts of the same temperature, their contraction will be uniform. There are castings that it is almost impossible to cool so as to keep straight, on account of their ill proportions; and should we by good management do so, we run the risk of having them crack sooner or later, as there is sure to be a strain somewhere, because a thin body of iron will contract in cooling more than a thick body.

With castings in which a strong artificial cooling treatment has to be resorted to, to make them come straight, it is best when possible to have the pattern made crooked to compensate for the warping when cooling. This is very often done in the case of castings that always warp one way. But there are castings in which this would be impossible, because they twist into all kind of shapes, one day being one way and the next day some other shape, and so on. About the best thing that can be done with such castings is to make them of iron, of the least possible contraction. If iron could be had that had no shrinkage qualities, we would not be troubled much with crooked castings.

Weighting down the ends or the middle of long, heavy, or light castings, is very often done to keep them straight. Sometimes the weights are sufficient to hold the casting straight without having to resort to any cooling process, and again it is necessary to weight them down and cool them as well. It is sometimes better, when possible, to weight a casting down than to try to straighten it by cooling. I have seen long awning castings go so crooked that it was necessary to take them out of the mould, lay each end on some pig iron and weight down the middle so as to bend it down more than an inch below a straight line, the bending being done the reverse of the way that it naturally crooked or warped. This was done as soon as possible after the casting was poured, so as to bend it while red hot. When the castings had cooled and the weights were taken off they came up straight.

Often moulders will weight down the ends of a casting,

and shovel the sand from the top in the middle in order to keep the ends of the casting from coming up. This is wrong, for by cooling the middle on the top side this part is contracted quicker than the under side, which will naturally cause the ends to come up if the weights are not heavy enough to hold them down. If the moulder is trying to hold down the ends of his casting, it would look more as if he understood what he was doing to dig underneath the middle of the casting so as to cool it, which, by causing the castings to contract there, will help the weights instead of working against them.

If the under side cannot be got at, it would be better to put on more weights, and instead of cooling the middle on top, leave it well covered over with sand, so as to keep it hot. It should always be remembered, that removing the sand from any part of a hot casting cools it ; and that cooling any portion of a hot casting causes it to contract the faster. Also that before one side can contract faster than another, there must be a crook or bend in the casting, this bend being made while some part of the casting is hotter than another. The part that cools first in equal-proportioned castings will generally keep the casting in the shape it was bent when hot.

It is not always that a moulder will succeed in straightening a casting before it is cold. He may, when he gets it out of the sand, see that it is crooked, and while there is yet a little heat in it, by dampening the rounding side or surface with cold water, straighten it. This treatment requires care, especially if the casting is so hot that you cannot bear your hands on it. The action of the water is sometimes visible as soon as it is applied, and if care is not taken, the casting will be bent the other way, in which case, if there is heat enough left, it may be wet on the other side to bring it back.

After castings are entirely cold, and are then found to be crooked, they may often be straightened by "pening" with a hammer, or by heating the middle with a wood and charcoal fire, and weighting down the ends.

The cut on p. 283 shows how this is generally done. The crooked casting is rested on two iron bearings, *E*, *E*, high enough to give a good chance to build a wood fire underneath. There are then wooden or iron guide-blocks, *P*, *P*, set to allow the ends of the casting, when it is being straightened, to be bent down the proper distance, so that, after the weights are taken off, the ends will spring up, leaving the casting straight. It will generally be necessary, if the middle is made red hot, to bend it down about twice as much as it is crooked; that is, if it is $\frac{3}{4}''$ crooked, it should be bent $1\frac{1}{2}''$, and after it is cold and the weights are taken off, it will be about straight. This is done by adjusting the blocks *P*, *P*, so that the weights will bend the ends down just as far as required.

In straightening castings in this way there is no cooling or contraction involved; the casting is simply bent, while red hot in the center, by the weights.

The casting must be red hot, and be bent more than sufficient to straighten it, as explained.

A wood fire is kept underneath, and a charcoal fire is made on top of the casting, so as to have it surrounded entirely by a strong fire.

Often castings can be made straight by "pening" the hollow side with a hammer, as shown. The hammering compresses and lengthens the part that is hammered, expanding the surface of the iron, which lowers the ends.

A solid iron block should always be used to rest the portion of the casting under the hammer, and the hammer should have a good, flat "pene." The hammering should be done gradually over the surface. Steady hammering on one spot and then on another breaks the skin or surface of the iron, or perhaps worse, will break the casting in pieces.

Page 283 shows how gates and runners sometimes draw castings crooked. This casting was a long, toothed rack, the teeth being small. The casting was gated as shown (sometimes such castings are run from one end, by having the mould set on an incline), and when it was taken out of the sand it was bent sideways. This was caused by the gates and runner being lighter than the casting, and in cooling and contracting the quickest, they draw the casting crooked.

To remedy this in the next casting, gates were made on both sides, as at *X*, *X*, instead of on one side only. This caused an equal pull on each side, thereby making a straight casting. The small gates, 2, 3, 4, and 5, at the ends, would generally be broken from the castings by their more rapid contraction.

There is hardly a foundry that has not been at some time troubled by castings bending or warping, and it is frequently troublesome to find the cause, and this often causes much experimenting before a remedy is found.

CAST IRON.

Cast iron is obtained from ores smelted in blast furnaces. The quality and grade of the iron produced depends upon the nature of the ores and fluxes used, also whether it is hot or cold blast, and what class of fuel is used to smelt them. The same class of ores will produce different grades of iron by varying the charges of fuel and fluxes; the fluxes are used to assist the melting of the iron, and to separate the earthy or non-metallic matter from the ores, which is run off in the form of a slag, similar to that which runs out of a cupola when large heats are run or dirty iron used.

The slag from a blast furnace generally contains more or less iron, and a well-managed and good working furnace is one that produces the grade of iron intended, and extracts all the iron possible from the ore.

Some ores contain more iron than others; about the lowest percentage of iron ore used is forty per cent., and about the highest percentage of iron said to be produced from ore is seventy per cent.

Silex, lime, and clay are more or less combined with iron ores; some ores contain so much that the ore will flux itself, and again one ore will be used to flux another.

"*Any substance which promotes the melting of another is called a flux.*"—Osborn.

The percentage and class of fluxes used depends upon the composition of the ores.

Sulphur, phosphorus, and manganese are properties that exist in the ores as in the cast iron, but the same percentage contained in the ore does not exist in the grades of iron obtained from them.

The manager of a blast furnace will sometimes cause the composition of ores to be entirely changed by his manipulation of charging and smelting them. For a manager to obtain the grade of iron wanted, a practical knowledge obtained from experience and from the chemical composition of the ores, fluxes, and fuels used is necessary.

The fuels used to smelt ores are charcoal, anthracite coal, bituminous coal, and coke. Charcoal makes the best iron because it is freer from sulphur, and other impurities which always exist more or less in coal and coke. Phosphorus in iron is obtained from the ores and limestone used to flux the ores.

Manganese in iron is obtained from the ores. The carbon and silicon in iron is obtained from the fuel used to smelt it with. The carbon in gray iron is mostly all in the form of a *graphite,* and the iron may contain as much as three or four percent. of it. A large percentage of graphite in gray iron will make it very soft, unless made hard by the presence of some hardening substance, such as sulphur, which, if not strong in the pig, may be derived from poor fuel or scrap in a cupola.

White iron contains carbon in a different state from gray iron. In white iron it is called *combined carbon,* in which form it hardens the iron. The graphite carbon in gray iron can have a large percentage made combined carbon, as in white iron by casting it on a chill or suddenly cooling it. By this action the carbon, which in melted iron is in the state of combination, does not have time to separate in the form of graphite.

Silicon, sulphur, phosphorus, and manganese are the constituents in iron, which, according as their percentages vary, regulate the grade, in making it strong or weak, hard or soft. Silicon and phosphorus greatly increase the fluidity of iron, thereby making it good iron to *run* light or thin castings. Silicon, up to 4 per cent, in iron increases its softness. Sulphur hardens iron, while phosphorus, up to 1 per cent, has a tendency to soften it. Manganese may soften or harden iron according as sulphur is present, but anything over 1 per cent rarely fails to harden it.

Charcoal is the only fuel now generally used with cold blast; anthracite coal and coke being blown with hot blast, and also charcoal, for the purpose of saving fuel and obtaining a larger percentage of iron from the ores.

White iron is sometimes produced by improper charging, bad fuel, or a furnace failing to work as it should. To obtain gray iron requires the most favorable conditions and management. White iron passes from the solid to a liquid state far more readily than gray iron. Gray iron becomes soft and pasty before becoming liquid, while white iron remains solid in its body to let the metal run freely from its surface in melting. No. 1 gray iron made from anthracite or coke furnaces is generally a soft, open-grained iron, and used to make thin, light castings; it has not much strength, but it possesses great softening qualities which permit an advantageous use of it mixed with scrap or harder grades of iron.

A No. 2 iron is harder, stronger, and closer grained than a No. 1 iron.

No. 3 is still harder than No. 2, and possesses much strength, while its color is gray inclining to white.

Nos. 4 and 5 are mottled irons which look like a mixture of gray and white irons.

No. 6 is white iron possessed of little strength, and is very seldom used except to mix with softer grades of iron in a foundry.

It is not always safe to rely upon the appearance of pig

iron (when broken) as to its qualities, for when it is melted in the cupola it will often return an entirely different grade of iron. This is due to conditions at the blast-furnace which cause parts or the whole of one cast to cool quicker than another, and thereby cause the same grade of iron to present different-grained fractures. The only way to correctly define the grade of pig iron before it is remelted is to have it analyzed so as to find out its chemical constituents. To judge of the merits of iron when in a liquid or solid form requires study and experience. Hard iron, when running out of a cupola, causes numerous sparks to fly in all directions; when in the ladle the surface presents an unbroken, close appearance, and if the surface is disturbed it acts sluggish and devoid of life.

A No. 1 iron in running displays no sparks, and when in the ladle its surface presents a lively, broken appearance of fine-colored undulatory movements. No. 2 and No. 3 present a similar appearance, only to a less degree as the iron becomes closer.

A piece of iron when broken, if good and strong, should present a medium-sized grain, of a lustrous dark gray color, fracture sharp to the touch, and close, compact texture.

A grain either very large or very small, a dull, earthy aspect and loose texture, indicates a poor grade of iron. The color of iron is generally lighter as the grain becomes closer.

Strong irons are inclined to hardness, weak irons to softness, in producing castings.

Cast iron melts at from 2,000° up to 3,000°, and will produce better iron when melted at a high temperature than at a low one.

MIXING AND MELTING IRON.

To be able to mix and melt irons that will answer for such castings as cylinder rolls, dies, pulleys, etc., requires experience and skill on the part of the founder, and he who accepts chemical analyses as a guide, instead of looking to the appearance of fractures, in judging the grade of pig iron, will be the most successful in making mixtures. Twenty-five or thirty years ago pig iron was not so easily procured as at the present day; much of the iron used was imported, and what few brands of iron were in the market were generally well known. At the present day, however, we find the home production so large that it would occupy pages to even mention the different kinds, and the importations are so small that we seldom hear of any. Most all American foundries now use different brands. Hence an attempt to give the names of the brands used would help the mixer or melter much less than the plan herein adopted. Different grades of iron have a higher or a lower temperature at which they will melt. A hard iron will generally melt faster than a soft iron. Pig iron requires a higher temperature to melt it as fast as a piece of old scrap iron of the same size. Pig iron generally melts at the ends of the pig first, for the sand and scale on them hinders the body of the pig from melting; and light iron will melt before heavy. When charging iron, the heaviest pieces should be put in during the first part of the heat. For cupolas under 30″ inside diameter, heavy pieces of iron should not be charged, as they are more or less liable to choke the cupola. In cupolas ranging from 30″ upwards, large lumps of scrap iron that cannot be broken are very often melted; I have charged pieces ranging from 100° up to 1,000°, but it is not economy to melt heavy pieces if they can be broken, for such large bodies require the use of more fuel to melt them. In charging iron it should be evenly distributed over the fuel, and made level, so as to receive the charge of fuel on the top of it. The question of close or open charging in effecting the combustion of coal or coke by the pressure of blast, is discussed on p. 308, Vol. II. In charging pig iron, the top smooth face, having no sand on it, should be the side to be placed next to the surface of the fuel, since, by so doing, the heat will reach it more readily. Long pieces of pig, also scrap iron, should be used as little as possible, especially in the smaller-sized cupolas, for they are apt to hang them up. In melting irons of different grades during the same heat, there is danger of getting them mixed, unless the greatest of care and judgment are used. Sometimes they will get mixed because of the method the melter has of charging, some men, when shoveling in coke or coal, will stand back seven or eight feet from the charging door, in order to avoid the heat. The fuel strikes the farthest side of the cupola, and lodges on the side nearest to him, so that instead of the fuel being level, it is banked up against one side ; the iron is now thrown in, and it rolls to the lowest side; thus the largest percentage of the fuel is on one side; and of the iron on the opposite side. There are cupolas charged in this manner very often. The foreman will complain to his melter because the cupola melts so slowly, or because it is choked before it has melted half the iron it should do; or when the castings come out, because they are hard, or not of the grade wanted; for any uneven charging of fuel and iron will always cause trouble, and to many it may appear a very profound problem to solve, when, in reality, it is only the lack of judgment and a little common sense.

SHOT IRON, OR BURNT IRON,

is a class of iron that foundrymen dislike to have anything to do with, on account of its mixing in and contaminating other irons and assisting to bung up a cupola. Shops that can use nothing but very soft iron in their castings generally have trouble to get rid of shot iron; there are some shops that will not bother with it at all; they will pick out all the iron they can from the cinder, and let the rest go; and it has often been a question in my mind whether they were not as well off as those that paid help to screen the cinders, and used the shot iron at the expense of good melting. I have tried in many ways to use shot iron so as not to retard good melting, and about the best plan is to make a separate heat of it, or melt it at the last of a heat for coarse work, or pour it into "pigs," which can then be used to mix in with good pigs in future heats; but in any form it should be used with caution.* Burnt iron, in some respects, is like shot iron, so far as making bad work is concerned. Burnt iron will make more slag than any iron, and cause a cupola to choke quicker. There are degrees in burnt iron, some are worse to deal with than others; but with shot irons it is all about alike. Burnt iron should only be used in small quantities at a time, unless a lot of sash weights, etc., are to be made, then a heat can be made of nothing but burnt iron, and the cupola run until your ladles and cupola are all choked, if one desires to do so.

* In either case it is of course charged, mixed with more or less pig and scrap.

IRON MIXTURES.

WHEN a foundry receives a pattern they receive instructions as to the grade of iron required in the casting. One may desire a good soft iron, another a very strong iron, or a very hard iron, or perhaps desire it chilled. These four elements comprise the requirements for special mixture. When a cheap iron is desired, the castings are generally made of what is called a common mix, and in such cases it may be even too COMMON.

Common mixtures are generally made of one third No. 1 soft pig, and two thirds common scrap.

Castings that need to be soft, as pulleys and thin castings, are generally made of the softest No. 1 pig that can be obtained, and then mixed with good soft machinery scrap iron.* If all pig iron is used, as a rule, softer casting will be obtained. By mixing two brands of No. 1 pig, in fact, to obtain any grade of castings from pig, it is best to use at least two brands of pig, as it results in a better iron. In the Eastern and Middle States more pig iron and less scrap is used than in the Western States. A great many places use almost all pig iron in their mixtures, which plan is not good in many cases, for a fair percentage of good scrap mixed with pig iron will often make a cleaner and stronger casting. Soft iron is not a strong iron; it is a difficult matter to obtain a very soft casting and at the same time a strong one. One of the best irons manufactured which accomplishes this end is the Hanging Rock iron, manufac-

* No. 1 pig iron is that containing sulphur not over .035 and having silicon ranging from 250 to 350. For a complete treatise on the effect of silicon, sulphur, and other metalloids, in changing the grade of irons, see "Metallurgy of Cast Iron."

tured along the Ohio River ; No. 1 Scotch pig is a good iron to use as a softener. When it is desirable to use up scrap or No. 2 and No. 3 pig, 100 pounds of No. 1 Scotch and 400 pounds of either will make a stronger mixture than if equal proportions of each are used. Nos. 1 and 2 of Scotch pig are generally very weak iron, and if used alone the castings may be porous and unclean when finished up. There is American Scotch pig as well as foreign.*

In making mixtures with any iron it is best to be always guided by chemical analyses. There are three classes of pig iron named by furnace managers—the red short, the cold short, and the neutral iron. The red short is an iron that has no strength when red ; hot the cold short is one that has no strength when cold; a neutral iron is made by mixing the red and cold short irons together, and, naturally, the neutral iron makes the best castings. When mixing irons of two or three distinct grades to make a casting which is to stand great pressure or strain, or to have a spotless finish, some plan to melt the mixture and pour it into sand pig beds; then, if the mixture gives the grade desired, remelt the pigs and pour the casting. This may often work well for some special cylinder, roll, or die castings, where a mixture of distinct grades is used, such as No. 1 charcoal and No. 1 anthracite or coke, car-wheel scrap, and a percentage of white iron or steel. Whenever white iron or steel is to be mixed with soft iron, the process is troublesome, for they will not mix well together; but by twice melting, the union is made more compact and the result cleaner. One thing that should not be lost sight of in remelting iron is, that every time iron is remelted IT IS MADE HARDER.

When making chilled castings a strong iron should be

* Silvery iron is now, in conjunction with Scotch iron, being used as a softener, and the same gives excellent results.

used, as the chilling of any iron will weaken it; the closer the grain of iron the deeper will the casting be chilled, hence a strictly No. 1 iron can seldom be chilled. Charcoal irons are generally used in making chilled castings on account of their superior strength. For castings that are to be chilled very deep or made very hard, white iron, and sometimes steel, is melted in with the charcoal irons. White iron makes a casting weak and brittle. Scrap steel can make a casting stronger and will harden softer irons.

For heavy castings, designed to stand strains and friction, two thirds No. 1 charcoal and one third mottled would answer. Mottled iron is generally a strong and close-grained iron. In mixing hard grades of iron, requiring finishing, No. 1 and Nos. 4 or 5 irons should not be mixed together, for it is apt to show an uneven grain in the finishing. The following receipts have been used for the castings noted:

LOCOMOTIVE CYLINDERS.

2,600 pounds of car-wheel scrap,
600 " soft pig.

These cylinders would be so hard that the edges and fins would often be chilled; the casting, when cleaned, weighed about 2,400 pounds.

MARINE AND STATIONARY CYLINDERS.

First—

One half No. 1 charcoal,
" good machinery scrap.

Second—

One third car-wheel scrap,
" good machinery scrap,
" No. 1 soft pig.

ROLLING MILL ROLLS.

Some places make their rolls out of car-wheel scrap only. For small rolls, however, if the rims of the car-wheels are not chilled over $\frac{3}{8}''$, and the middle and hub appear tolerably soft and not mottled, the iron would be the right grade. The wheels selected to make rolls over 14″ in diameter should be the thickest chilled ones, and the rolls so made often have the edges and fins chilled.

One half car-wheel scrap,
One quarter No. 1 charcoal,
" No. 2 "

This is also a mixture used for making rolls, and the car-wheels should be selected for the small and large rolls as above noted.

Mixture used for making small chilled rolls, which were desired chilled $1\frac{1}{2}''$, otherwise they would be of no use ·

1,300 pounds of	old car-wheels,
100	No. 1 charcoal,
300	steel rail butts.

Mixture used for making kettles which had to stand a red-hot heat all the time, so that the iron had to be strong and close ·

1,300	pounds of	No. 1 charcoal pig,
800	"	old car-wheel scrap,
700	"	good machinery scrap.

Mixture used to make castings chilled, which are moulded all together in sand, the castings being required to stand friction and no strain :

200 pounds of white iron,
200 " plow points,
100 No. 2 charcoal,
100 car-wheel scrap.

PULLEY MIXTURES.

Iron for making pulleys should have as little shrinkage about it as possible. It would be a hard matter to give the exact proportions of iron to be mixed for them, as the thickness of the rims and the quality of the iron is what such mixtures depend upon. For thin pulleys the iron cannot be mixed too soft, sometimes it is best to select the openest pigs from two brands of a No. 1, and again two thirds of No. 1 and one third good scrap work well. For rims over $\frac{3}{8}''$ thick, often pulleys are easily turned up, if the mixture is of equal proportions of No. 1 and good scrap.*

SASH-WEIGHT MIXTURE.

Two thirds scrap tin,
One third stove plate scrap.

This mixture, when melted, made white iron.

The few mixtures given show how special grades can be made or changed. To mix the above iron for machinery castings, these mixtures, except the sash-weights, are generally costly, and always demand an increase in the price of castings.

---

* A high silicon and low combined carbon pig iron, known as "silvery iron," is now used to mix with No. 2 grades of pig and scrap iron for producing soft castings. *High silicon pig must be used carefully, for if it exceed* 3.00 *of silicon in castings it makes them weak and brittle.* Up to 4.00 in castings, silicon will soften iron, but in excess of this amount it will harden it.

ODD WAYS OF MELTING IRON.

THERE is probably as much reason for changes in the plan of melting iron as there is in moulding jobbing work. Melters will sometimes get nervous at being ordered to charge up their cupola in as many different ways as there are days in the week. A foreman that understands his business very seldom lays out a system, or a table of charges, for his melter to follow day after day, in a regular jobbing shop. The foreman may have various reasons for wanting his melter to make all these changes. To-day he may want the cupola to melt extra fast during the first of the heat, and slowly after some heavy casting is poured, in order to have melted iron to feed with. To-morrow, seeing that some moulder will not get ready in season, this order may be reversed. As he does not want to keep his men late when it can be avoided, he orders the cupola charged, so that the men hav ing small work can be pouring off while the large casting is being got ready. This casting, that, perhaps, weighs five tons, may not be thick in any of its parts, so as to require much feeding, and the bottom can be dropped soon after it is poured. In this way the only moulders kept late are the ones that were going to keep the whole shop's crew behind, which, for a shop that pays overtime, would be expensive, and, in any case, is not pleasant for the men.

On some days the shop floor may be covered with a class of work that is better for being poured with dull iron, and the next day the work may be such as to require very hot iron. Again, there will be heavy and light castings, requiring entirely different grades of iron; and to complicate matters the foreman, if an observing man, will see that the brand of iron is not of the same grade as the last car load. All of the above causes, to which could be added quality of fuel, sometimes make a thoughtful foreman think of a string tied full of knots.

I will try and show in this, two of the many plans that may be adopted to meet different conditions that may be new to some. One is for melting special grades of iron, and the other to retain the bed in a cupola after melting a heat for a break-down job, or for a piece of casting that is wanted in a hurry.

I worked once in a rolling mill company's foundry, and sometimes when everything was about poured off, there could in the distance be seen some one of the managers running towards the foundry as if he meant business. Our ignorance of the cause of his haste would soon be enlightened by seeing a team, or some men bringing a pattern. This pattern would be given to some competent moulder, and two or three reliable moulders would be retained to help him. By this time all the moulds are poured off and the cupola man has received instruction not to drop the bottom, but to prepare it to melt iron again in the course of three or four hours. The way to do this is as follows: Leave the blast on until you are sure all the iron in the cupola is melted, and instead of dropping the bottom knock out the front breast, and with a bent hook pull out all the clinkering coke or coal and iron cinder that can be felt or seen. Then fill up the breast hole with loose sand, and every five or ten minutes take away the sand and pull out again whatever clinkers or iron cinders will have formed, repeating the operation for the first half hour or so, or until you are sure that all the droppings of iron and clinkers are pulled out. After this, every half hour or so will be sufficiently often to clean the bottom out. The stopping up of the breast every time the clinkers are cleaned out is done to prevent the fuel from burning away, and also to keep the clinkers and droppings of iron from being chilled with the air.

After the cupola is well cleaned out, there should be some fuel shoveled in, so as to freshen up and keep the fire in good burning condition. When the moulders have their mould or moulds about ready, then make up the breast as usual, and shovel in the fuel for a bed, the same height as for a regular heat. After it gets to burning, charge up the iron wanted, put on the blast, and you will soon have your cupola melting iron again. The first two or three hundred of iron is generally dull, and sometimes will have to be poured into a pig bed. After this the iron will come hot enough for ordinary castings.

The question of how large a heat a cupola run in this way would melt could not be better answered than by the following: One morning early, two or three men were called upon to mould up a piece of machinery for a repair job. The melter and helpers were called to the shop to get the cupola ready as soon as possible. The casting, the weight of which was about 2,500 pounds, was poured about eleven o'clock in the morning. The iron was all blown down, the breast knocked out, and the cupola treated as above described, until the time for the regular afternoon heats, which were never less than 12 tons. The blast was again put on, and after the first few hundred pounds the iron was as good and as hot as usual. The time that the cupola was held from one heat to another was about four hours. The size of this cupola was a five-foot shell.

To prevent the mixing of different grades of iron, when melted at one heat, has been the cause of a deal of thought and many experiments with foundrymen. I know of a foundry owner who makes a practice of melting only one grade of iron at a time. If he has a roll to cast, he will only charge up the iron weighed off for it. The blast will then be put on and all the iron in the cupola melted and tapped out. The blast is then stopped and the bed renewed with coke. Another grade of iron is then charged up and all melted down. I remember one day he made three distinct blow-outs during the same heat. The first was about 7,000 pounds for a roll; the second, about 2,000 for soft work, and the third was common iron to finish off a heat of about 8 tons. The size of the cupola that this was done in was about a four foot six inch shell. This same gentleman has a reputation for turning out castings of the grade of iron wanted, and it is owing to no more nor less than the way he charges up his cupola, and in being particular in the mixing and selection of his iron. The objection to this style of melting is that there is a little more coke used, and it takes from half an hour to one hour longer to run a heat off.

It seems almost an impossibility to run a straight heat, when there are two or three different grades of iron to melt, without having them mix more or less, and the less the weights of the different grades to be melted, the more will they be liable to mix. For example: Charge an ordinary cupola with a regular charge of a special grade of iron, with the usual charge of fuel on top, and so on, charging with distinct grades of iron. As the grades of iron melt, pour some castings, the weight of which should be nearly the same as the charge. On the following day melt the special grades of iron by themselves, and pour some castings, and then compare the runners and gates, and you will see that there is a difference.

It is generally known that hard iron will melt sooner than soft iron, and most foundrymen, when making a casting of hard iron, have the hard iron charged first, to make sure of having the casting of good, sound iron and of the grade wanted. If they have soft iron to run, it is generally charged on the top of the hard iron. This is a plan that I do not always approve of, as there are always more or less particles of any grades or charges of iron left remaining among the fuel and on the bottom and sides of a cupola, and which will affect two or three other charges.

A plan that I find to work well, when hard and soft iron are wanted, is to melt the hard iron first; then, instead of putting the soft iron directly on the top of the hard iron, I charge one or two charges of common iron. On top of these charges the soft iron will be charged. After, as I think, all the hard iron is down, then the common iron is tapped out until, by the number of ladles carried off, I think it is all melted. At this point the soft castings are poured according to the degree of softness wanted. The softest casting wanted, if there have been three charges of soft iron charged, should be taken from what is thought to be the middle or second charge.

In some cases where I have only a small amount of very soft iron wanted, I charge up the soft iron on the top of the bed, which should be burning well, and should not have in as much fuel by from 4″ to 6″ as for ordinary heats. This iron will be put in from one half hour to one hour before any of the other charges of iron are put in, and when all is ready to have the rest of the charges put in, make the first charge of fuel (that which is placed between the first and second charges of iron) a large one; as much larger than usual as the bed was left low. By this means the large charge of fuel takes a longer time to get hot, and separates the charges of iron more readily. When the first charge of iron is melted, the second, or large charge of fuel, will come down and raise the bed up to the proper height to run the balance of the heat off. I have by this plan charged hard iron on the top of soft iron.

And when not taking out the soft iron too closely to the amount charged up, the castings have been as soft as if the hard iron had never been charged up. It is in having only small quantities of different grades of iron to melt that there is serious trouble with their mixing together. With large quantities there is more chance of having castings the grade wanted; but even then the melter must use judgment in seeing that the iron is charged as it should be, and the foreman should be watchful, so as to know that the iron is taken away from the cupola as the grades melt or come down.

THE TUYERES AND LINING OF A CUPOLA.

THE governors of a cupola are its tuyeres: it is through them that life and combustion is given to the fuel by rapidly supplying air. Without air there can be no fire, for the oxygen air contains, when combined with the carbon in the fuel and ignited, gives to us heat or flame, so that the more we supply this oxygen to the fire, the faster is the fuel consumed. Chemists tell us that two atoms of oxygen combined with one atom of carbon cause a thorough combustion of the fuel, and if more than two atoms of oxygen are supplied to one of carbon, it causes a destruction of the fuel by making its life short.* To obtain the heat for the hot and fast melting generally required, our forced blast of air is said to give us more than the two atoms of oxygen, and hence we are compelled to use more fuel than we otherwise should. There are manufacturers of patent cupolas who claim their process will largely prevent this extra consumption or waste of fuel owing to certain arrangement of the tuyeres, and among the most prominent are the *Colliau* and *McKenzie* cupolas. The best test of these patents is their practical working, which must be seen to be understood. The *Colliau* cupola tuyeres are based upon a very scientific principle to accomplish the end desired: since, however, it is only intended to notice the various cupolas and tuyeres which may be used, no recommendation is made. There are every imaginable shaped tuyeres used,—oblong, triangular, oval, square, flat, and round. For each style there can be found ready advocates; but, after all, the plain round tuyere has as many points in its favor as any style used. *The form of a tuyere is of*

* A lengthy discussion upon the question of "Blast and Combustion" is given on p. 305, Vol. II.

second importance to the question of their proper distribution and area, for which see pp. 301–320, Vol. II.

The distance of a tuyere from the bottom or bed is determined by the class of work to be done; for instance, in foundries for making stove plates, the height of tuyere from the bed should be from 7″ to 15″; while in machine or jobbing foundries they should be higher, say, from one to three feet, according to the amount of iron required to be melted at one tap. The advantage of low tuyeres is a saving of fuel.

For melting large quantities of iron, it requires the same amount of fuel over a low tuyere as it does over a high tuyere.

Another reason for having high tuyeres for use in machine or jobbing foundries is, a large body of iron is often required to be melted before tapping out the iron into a "crane ladle." The object is to have a large body of iron to retain the heat, as sometimes it takes two or three hours to melt enough iron to pour a heavy casting. This course also gives time to allow the scrap iron of all descriptions and grades, also heavy solid pieces of old castings, to melt and become thoroughly mixed with the new iron which has been added.

A cupola with tuyeres high will melt more and run longer heats than it would if the tuyeres were low; but there are times when having both would be an advantage. To meet this want, there have been two sets of tuyeres applied to the cupola, and placed one above the other. These can be easily arranged, so that either set may be employed to ad vantage, using the high tuyeres for heavy heats, and the low tuyeres for light heats.*

The openings of the tuyeres not in use are to be stopped with clay. Sometimes the spout and breast of a cupola can be so arranged as to raise or lower it, thus affording an opportunity to put in a high or low sand bottom, a plan

---

* Another point which goes to regulate the height of tuyeres is given on p. 276, Vol. II.

which not long ago was used by the author in a 50″ cupola, and found to work satisfactorily. A very convenient form of alarm for indicating the highest limit to which the melted iron is allowed to rise in the cupola will be readily understood by the following description. Referring to the aecompanying engraving, it will be observed that the melted iron has reached the highest limit allowable, and is running through the tuyere hole into a small cast-iron box, having an inclined wooden bottom.

This bottom, shown at *X*, has three holes of one inch diameter, bored through within $\frac{1}{8}''$, allowing sufficient material to prevent the wind from escaping.

The bottom is held up tightly in place by a piece of round iron and a wooden wedge, as shown. This device should be attached to the tuyere nearest the spout *F*, so as to be easily observed. It is of essential importance to have the tuyere, to which the alarm is to be attached, about one inch lower than the rest, in order that the alarm may be given in time to prevent the melted iron from running out of the higher tuyeres into the pipes *G*, *D*. When the melted metal rises to the height of the low tuyere, it will run into the alarm box, filling the holes and burning through the wooden bottom to the floor almost instantly. A ladle could be placed under the alarm to catch the melted iron, if desired, without doing any injury. Several extra wooden bottoms should be kept on hand to replace those burnt out.

The tuyere valve, *B*, forms a very convenient air-tight opening, and furnishes the means to bar into the cupola, or inspect the same, as a piece of mica is fastened into the opening *e*, with putty.

The application of the alarm described, to a cupola, effectually prevents the excitement which usually prevails in a foundry when the melted iron overflows, resulting in heavy losses of castings.

The workmen are frequently and sometimes badly burned by accidents of this kind; and there are many cupolas in use having quantities of iron in the wind boxes and pipes, thus obstructing the passages. The pipes are frequently destroyed by the hot metal, while in others they have to be patched. The workmen have to rely upon their judgment generally to determine the height of the iron in the cupola, and sometimes are deceived. In some instances the tuyeres are so constructed that an alarm could not be applied to them; in such cases I would recommend the application of a blind tuyere one inch lower than the working tuyeres, and attach the alarm to it. This useful appliance is, I believe, original, and is hereby given to those who may wish to use it

The lining of a cupola should always be built solid and close. The fire-clay placed between the bricks is only to make an air-tight joint, and the less clay used the better. The clay should be mixed with water, and very thin, so that by dipping the bricks into it sufficient clay will adhere to them to form a tight joint. Each brick should be hammered until all the superfluous clay is squeezed out from the joint. A cupola lined up in the manner described will last one third longer than when the bricks are laid in thick clay, keeping the bricks apart; and as the clay has not the power to resist the intense heat, it soon crumbles away, leaving the joints exposed to the action of the fire. In mixing clay, some advocate the addition of one third sharp sand. A very good plan is to boil the clay and sand together in a pot, as they will become more thoroughly incorporated. There are three courses or thicknesses of bricks used in lining up a cupola. Some foundrymen line up their cupolas with a four-inch wall, keeping the bricks back from half an inch to one inch from the shell of the cupola, filling the open space with clay, making a wall of about five inches thick. When this course is adopted, the man in charge need not be surprised some day,

Lining

Lining

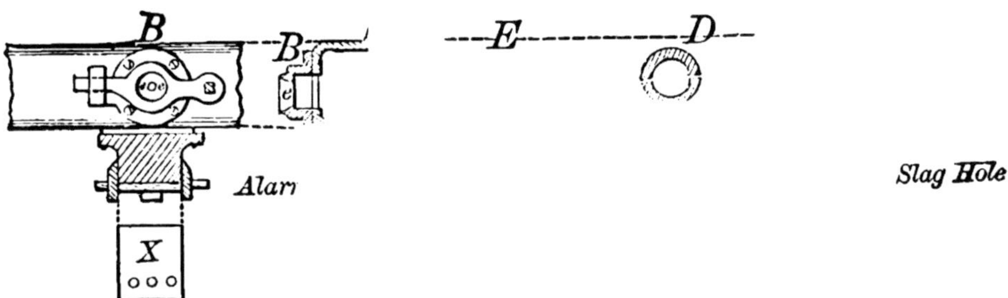

AN ALARM FOUNDRY CUPOLA.

after the bricks have burned out a little and a heavy heat is raised, to see the cupola shell get red hot, and perhaps a hole burned in it. I claim that the safest and best way to line a large cupola is with an eight-inch wall, as with that thickness of bricks no fears need be entertained in running off heavy heats, and when the inside four-inch lining is burnt out, it may be replaced without disturbing the other four-inch lining next to the shell.*

When it is impossible to follow the above directions, on account of the double lining making the inside diameter of the cupola too small, very satisfactory results may be obtained by placing the bricks on end, so as to make the back lining two inches, and a four-inch lining laid up in front, as shown by the engraving. Broken bricks may be used for the back lining. The dimensions of fire bricks here given are more theoretical than practical, for some will be 4½″ wide, while others may be 4¾″ or 5″, and vary in thickness from 2¼″ to 3″.

The inside lining in this case can also be replaced without disturbing the 2″ lining. The inside lining should not be allowed to run too long before replacing, because when it is allowed to go beyond a certain limit, pieces of the bricks will flake off, mixing with the melted iron, forming an excess of slag, causing a retarding of the melting process, and producing dirty castings. The destruction of the cupola goes on more rapidly under the conditions named.

A slag hole should be applied to cupolas in machine foundries, as it is very essential in keeping the cupola clean and forwarding the melting when using dirty or burned scrap iron, or bad fuel. Any of the foregoing substances would tend to make plenty of slag, particularly during a large heat. Even with a small heat there is more or less formed; and there are many cupolas where the slagging is all done through the tapping hole, which is a very dirty process, besides burning up the ladles when there is much slag. The

* Common hard red brick can, in the place of fire bricks, be used to form the 4″ lining next to the shell, but of course they are not so reliable.

proper place for a slag hole is behind the cupola, because it is out of the way. It should be located from two to four inches below the tuyeres. When slag is forming, and it is desired to let it out, the cupola should not be tapped until the slag has reached the level of the slag hole; the hole may then be opened and the slag allowed to run until the iron appears, when the hole should be stopped.

The spout should then be tapped, and from 300 lbs. to 800 lbs. of iron allowed to run out according to the size of the cupola. The iron should then be stopped, and in a few minutes the slag hole should again be opened, after which from one to five tons of iron may be melted without the necessity of opening the slag hole again.

During some heats it becomes necessary to slag out several times—depending upon circumstances. A slag hole should not be located directly beneath a tuyere, as the blast would drive the slag back, preventing it from coming out. I believe that if foundrymen who have been accustomed to slag out at the tapping hole would adopt the plan of a separate slag hole, they would be so pleased that they would never think of returning to their old methods.*

* To thoroughly understand the benefit derived from "fluxing" and "slagging out," and to obtain further information in manipulating the same, read chapter on "Slagging out Cupolas," p. 310, Vol. II.

PREPARING CUPOLAS.

THE various odd shapes given to foundry cupolas are generally the result of circumstances.

There are traditions extant of men, who, in commencing business, could not afford a cupola possessing the proper qualities and improvements; so barrels or tanks were lined with bricks and clay by some, while others, who were more enterprising, made a square cupola of open sand-plate castings bolted together. These make-shifts will do for past generations and in localities where there is a lack of capital. But the business man who understands how to run a foundry economically, insists upon having a first-class cupola, if it is to be had. There are two principal styles of cupolas, viz., the oblong and round.

The former possesses the advantage of allowing whole pigs and long pieces of iron to be "charged up," without requiring them to be broken in small pieces; the latter style is, however, more generally used. Cupolas can be used from 10″ to 72″, or even larger if desirable; small-sized cupolas are generally made with swivels, for the purpose of dumping them when they have melted their small heat. The small cupolas are only practicable for melting small quantities of iron, as, for instance, casting some light job or testing new brands of iron.* To run a foundry with the intention of making money, no one should start with a cupola less than 20″. The common sizes of cupolas range from 30″ up to 48″ (these measurements are inside diameters). The amount of iron a cupola will melt depends greatly upon the man-

* An original and good plan for constructing small cupolas is illustrated on p. 265, Vol. II.

agement, the character of iron and fuel used, and whether it is "slagged." Improvements on the inside of cupolas have been attempted in various directions, but thus far the common straight cupola, as shown in the cut, has not been improved on.*

I have often thought that the simpler the construction of a cupola the better will be the results, and the longer I live the more I believe this to be true. Just take a good look at the inside of a choked cupola, and then think how long and how much work it would require to keep any portion of it in an octagonal, hexagonal, or any analogous shape, and I think you will conclude that such forms were not desirable. A plain, round, straight-lined cupola, made with the bottom larger than the top portion, is the best for cupolas under 30″. Above 30″ there will not be any trouble from having the bottom and top of the same diameter; and, to my mind, a cupola should not be smaller at the tuyere, unless more than 48″ in diameter—inside measurement.

In small cupolas there is generally difficulty in respect to choking, which occurs when the cold blast has not a sufficient quantity of live fuel to make it hot in reaching the center, and also from the liability of the pig and scrap iron becoming fast in its downward flight; by making these small cupolas larger by 3″ or 4″ at the bottom than at the charging door, the iron and fuel become looser as they descend. Larger-sized cupolas, made the smallest at the tuyeres, upon the plan of the McKenzie cupola, generally give good results, for there is some fuel saved, and the blast brought with more force into the center of the cupola or fuel.

The height to make cupolas ranges from 7 to 14 feet, the height increasing as the diameter is enlarged. High cupolas confine or hold the heat, and make the iron hotter, and melt it faster when it gets down to the melting point than

* For a well-defined table upon the capacity of cupolas ranging from 20″ up to 80″, see p. 314, Vol. II.

low cupolas. The number and size of tuyeres the cupola should have depends somewhat upon the shape and construction of the tuyeres. With a plain round tuyere

A 20″	cupola	can	have	two	5″	tuyeres
A 30″	"	"	"	three	5½″	tuyeres
A 40″	"	"	"	five	5½″	tuyeres
A 48″	"	"	"	seven	5½″	tuyeres,

all evenly divided around the cupola.

For size of the main pipe which carries the blast from the fan or blower to the cupola, see table on p. 318, Vol. II. All right-angled turns in blast pipes should be avoided, as they break the force of the blast. For definite information upon the sizes of tuyeres, height, etc., best to use with coal or coke in different sized cupolas, see pp. 301–320, Vol. II.

Good management in melting iron is only indicated to the observer by the amount or weight of iron or fuel used in charging up a cupola, and the time consumed in melting. This information is good as far as it goes. A man knowing this much, if he had the cupola prepared, could charge it up and melt iron, and have a reasonable success so long as the grade of iron, fuel, and working conditions did not change.

It is more difficult to prepare a cupola properly than is popularly thought. The first thing a cupola man generally does in the morning is to put away his ladles and shanks, etc., and if he has any helpers they may assist him, or be gathering the scraps and gates. Some places will "jingle" their small gates so as to cause the cupola to melt faster and cleaner. After all the tools are put by, the melter will be getting the ladles ready, while the helpers are getting the cinders away from the cupola, and mixing the clay for him.

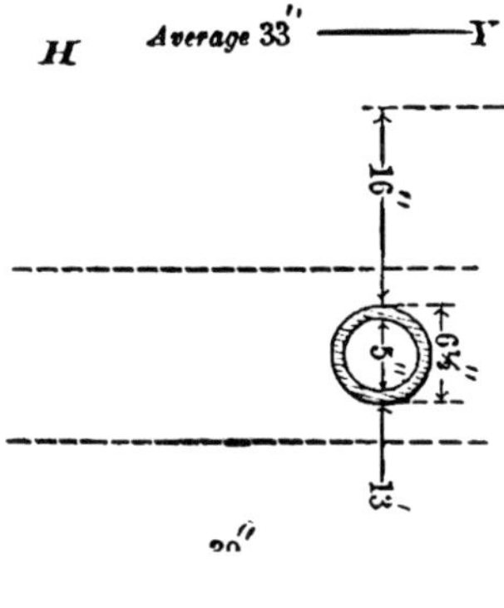

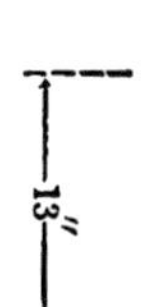

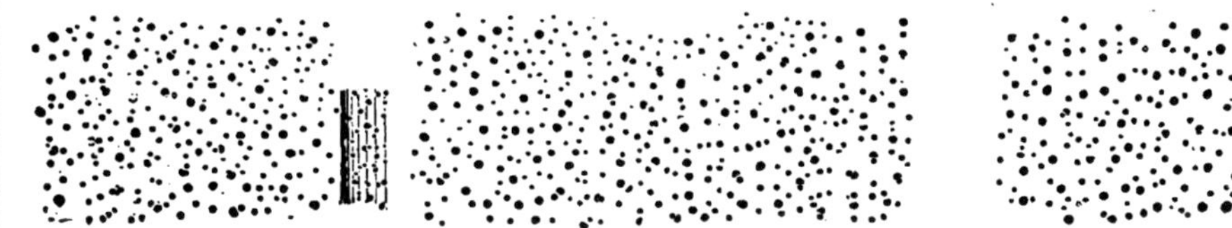

CUPOLA WITH MELTED PORTION DAUBED UP.

Next the melter will go inside the cupola and pick it out with a small, sharp pick, being careful not to break or disturb the face of the bricks, for if they can be left with the thin, glossy skin or cinder formed upon the surface of the brick by the use of the fluxes and from the heat, it will often stand the fire and blast better than some of the clays used. After the cupola is picked out it must be daubed, and in doing this many melters think the melting point, or that part of the inside that gets burned out, should be filled up so as to be level with the rest of the inside; but this should never be done.

In looking at the cut shown herewith, there is seen one side of the melting portion daubed up, so as to fill up all this burned or melting surface even with the upper and lower parts. This is one cause for cupolas getting choked before they have run half of the heat off that they would if daubed up as shown on the opposite side at *Y*. Too much clay daubed on the lining will only bag down, as shown at *H*, and make it too heavy to hang on the lining or bricks. When the blast is put on, the commotion of the fuel and iron against it can soon start and cause it to come away from the lining and mix in with the fuel and iron so as to cause a large amount of slag.* It will also form a bridge over the tuyeres, thereby preventing the blast from getting into or among the fuel. When this daubing falls off, the iron and fuel will get in over the bridge, and cool and chill so as to soon bung up the cupola, and stop the melting. To properly daub a cupola there should not be more than one inch of clay on any part, and when the lining is burned out in spots there should be some pieces of fire-brick built in with the clay, to save having large lumps of wet clay to dry and to cause trouble when the blast goes on. The melter should be very careful in mixing his clay to have just the right quantity of open or sharp sand mixed in with it, as too little

* The steam generated from the damp clay in trying to relieve its pressure, is also a factor to disturb the daubing from the cupola sides.

causes the clay when drying to crack open, and too much destroys the body of the clay.

The best clay for daubing is fire clay. When the melting point or surface is so badly burned out that from ½ to 1 inch thickness of clay will not keep the iron shell from getting red hot, that portion of the cupola should be relined at once.

The lining of a cupola will last twice as long where good fire-clay is used in preference to the common clays so frequently used. It costs more to purchase fire-clay than the common red or blue clays; but as to the question which is the cheapest in the end, it may be noted how very costly it is to reline a cupola every few months. When a melter in picking out the cupola sees that it is burnt out in some places more than in others, he may be sure that something is wrong; either the charges have not been put in evenly, or the cupola has not been daubed properly. These two things are sure to cause uneven melting. After the cupola is daubed, put up the bottom; this is generally done by propping up a section or one half of the door, and then shoveling up the sand, after which the balance of the door or bottom is permanently propped up as shown. Very large cupolas sometimes have the drop-door in four sections, medium-sized cupolas in two, and in those under 30″ the drop-door is made all in one piece and hung by two hinges. Drop-doors are sometimes made of wrought or boiler iron, so as to make them lighter; cast-iron ones being too heavy to put up easily.

The sand used to make the bottom with is picked up from the gangways, or from dirt piles; sand for forming bottoms should not be too loamy, for it would be apt to bake hard, and not allow a bottom to drop, especially in small-sized cupolas. Many a man has been burned in endeavoring to pry down baked bottoms. It is also bad to use rotten or very open sand, because the iron is apt to wash it away. It is advisable, after the sand is rammed down and the shape of the bottom formed, to coat it with clay wash, for by so doing a firm crust will form on the surface. The sand should not be too wet, or rammed too hard, as either will cause trouble, just as the bottom of a green sand mould does from wet sand or hard ramming. The author has seen a cupola bottom blowing so that the sand was lifted enough to let the iron run out at the bottom. A bottom should be made sloping, as shown in cut; this is to make it certain that the iron will all run out. If a bottom has too much slope it will cause the iron to rush out with force, and hence make it difficult to stop it; while if there is not enough slope the iron is apt to choke up at the breast or tapping hole when it first commences to melt. Putting in the front or breast in a cupola should always be done intelligently, or there will be a failure of some kind. The front of cupolas is made large enough to admit shoveling the sand through them to make the bottom with, if desirable. When the lining of a cupola is over 6″ the brick had better be cut away from around the front, so as to form the tapping hole of a proper length; a long tapping hole will always be troublesome if the iron chills in it, and also it makes an ugly-looking front. A tapping hole should not be over 3″ long, and made with clay, so that the working of the tapping bar and washing of the iron will not wear it away. The front or breast can be rammed with a mixture of clay and new moulding sand, or let the whole front, including the tapping hole, be formed from a stiff clay. Some melters do not put in the front until the fire is started, using the fuel for a backing to ram against; others will make it half up before putting in the fuel, and then, after the fire gets burning nicely, they will put in their draw plug and make up the balance. A good plan is, when the cupola is large enough, to have a board with a hole to admit the tapping draw plug held up against the inside, while one rams or packs up the front until solid, after which, with a trowel, make the inside of the hole of a conical shape, as shown, and make the clay smooth and even with the brick-work; also it is well to have some clay in place of sand to form the bottom for 4″ or 5″ beyond the inside of the breast, so as to prevent the tapping bar from making a hole in the bottom. For small cupolas sometimes a piece of a board set in against the hot fuel is used to form a backing to ram or pack the breast clay against. The spout of a cupola is made while the breast is being formed, and dried a little with charcoal, or some hot coals, before the cupola is charged. In preparing a cupola most every cupola man has some method of his own, for it is a branch of the moulder's trade that men have generally been left to manage or pick up by themselves as best they may.

FUEL AND CHARGING IRON.

To melt iron we must use fuel, and by the quality and class of fuel used the nature of the iron is more or less changed. Fuel that contains an unusual amount of sulphur will always make the iron hard, and also effect quality of slag. A good method of testing fuel before using is to make a piece red hot, and let it drop into a pail of water; then by practice it is possible to tell by the smell if it contains an unusual amount of sulphur. Coal generally makes a purer and softer casting than coke. The percentage of fuel required to melt iron depends upon the height of tuyeres, pressure of blast, and the quality and grade of iron used, as well as on the construction of the cupola, and the quality of the fuel. With coal or coke that is hard and clear less fuel is required than if it is soft and flaky. It requires more fuel to melt heavy iron than to melt light, and all pig iron requires more fuel than heavy scrap on account of the sand on the pigs. The stronger the blast the higher it is forced through the hot fuel before it becomes heated by its union with the fuel to its greatest temperature. When there is too much fuel for a bed we do not obtain the full benefit of the best melting point, as the iron is raised up above it, and since it melts above this point it will not melt so fast. The same is true when the bed or fuel is lower than this point, and it is worse to have a low bed than a high one, for a low bed will cause dull iron. For the iron, when it is melted, is not obliged to drop through the highest point of temperature, as iron does when melted upon the high bed, and further, the blast is colder in the low bed than in a high one. A workman, to do RAPID or ECONOMICAL melting, should vary the height of his bed until he gets the iron hot enough and melting in the fastest time. He should also be careful in the management of the cupola, and particular in charging it. There are hardly two cupolas that will be found to have the same blast pressure. Attached to the last cupola is a water glass tube inserted in the wind box, with a cork on its end to prevent its breaking. A tube thus applied and filled with water will show in inches the pressure of the blast on the outside of the cupola; but to determine the inside pressure is a hard matter, for the inside diameter of the cupola and size of the tuyeres will always cause the pressure to vary. It is not always true when the pressure is high on the outside that it is the same on the inside. There will generally be the most pressure shown toward the latter end of a heat, and this is caused by the tuyeres and fuel in the front of them being choked more as the melting increases. The pressure of blasts used will be found to vary from eight to twenty inches. It requires a stronger blast to melt iron with coal than with coke. A weak blast will cause slow melting, and too strong a blast is apt to harden the iron and make slag, since its power will cut the clay and lining. Coke will melt iron faster than coal, and a cupola should melt longer with coal than with coke.*

Coke and coal are often used together. In sections where coal is the more expensive of the two, some foundries make a practice of filling up to the bottom of the tuyeres with coke, and then making the balance of the bed all coal; and between the charges of iron they will throw in from one fourth to onethird the amount of coal to that of coke, putting the coal in first, which is a good plan to adopt in coke sections when there is a very large heat required. Again,

* The particular merits of coal and coke are more fully defined on p. 273, Vol. II.

some will use all coke for the bed, and use a little coal between the charges, and others will have the order changed, using some coal on the bed and none between the charges. The height of fuel required to form the bed is lower for coal than coke; eighteen inches is about the average height for coke, and twelve inches for coal.* Above the top of the tuyeres medium-sized fuel will give better results than large lumps of coal or coke. The medium fuel makes a hotter and more compact fire. Imperfectly started fires have often caused many bad castings. Melters are often seen putting on a weak blast to make their fire burn up, so that they can commence charging up their iron, and oftentimes melters cannot obtain dry kindling wood enough to properly start the fuel. But as a general thing the melter is to blame for the careless manner in which he goes about his work. Sometimes he will not take the pains to split up planks or timber as small as they should be; again, he will not have enough shavings to properly start his wood, or he may have a lot of short pieces of wood or blocks, and he will put them into his cupola in such a manner that a stranger would think he was trying to see how small a space he could pack them into; or again, he will not have wood enough to get the fuel properly lighted, or the fuel will only be burning on one side. To properly start a fire a good melter always tries to have a well-dried supply of kindlings on hand, and not wait until he wants to use it, and then take the first thing he comes across, even if it is wet. Kindling wood is effective when split up in long strips, and placed endways on a slant against the side of the cupola so as to protect the daubing as well as to catch fire better; and if small pieces are used, let them be laid in the middle as open as possible, and on top of this kindling do not place any more fuel than is necessary to kindle additional fuel after the wood is all burnt out. Too much fuel at first

* This means, above the top of tuyeres.

has put many fires out, or made them burn poorly. If a fire does not kindle at first, there is often dull iron all through the heat, in spite of all efforts to overcome it. The proper time to start a fire depends upon the class of fuel used. A hard coke or coal fire should be started sooner than a light or porous fuel fire, for the hard fuel requires longer in order to get it started properly; but if the same time be given with softer fuel, much of its life would be burnt out before the blast was put on. A fire should be started soon enough to get a cupola well heated up before any iron is charged. About the average time of starting fires is about two hours before the iron is charged, and the iron is better to be charged about a half hour before the blast goes on, as by so doing it will get heated and melt faster. Upon the bed of fuel in a cupola is where iron is melted, and the height of this bed should be kept even with the melting point until the latter end of the heat, when it can be allowed to become lower. To keep up this bed there is fuel put between the charges of iron, and as they melt the fuel comes down on the sinking bed and raises it up to its proper point. By the amount of fuel charged between the charges of iron the character of the melting will be much regulated. To make even and regular melting throughout a heat the melter should know what percentage of fuel it requires to melt a hundred pounds of iron when charging, in order to replenish the consumption of fuel on the bed. To know the proper percentage to use, the melter must rely on experience as practiced in his own cupola.

The weight of iron to use in making charges generally depends upon the class of fuel used, and on the diameter of the cupola. With coal, charges need to be made larger than with coke. With fifty pounds of coke between five hundred pound charges of iron, in the size of the cupola shown, we have enough fuel to cover the iron over and separate one charge from another. But were the charges thus made with coal, the coal would not separate the charges, and the iron would appear as if it was one solid body, from the charging door down to the bed. So that in order to successfully melt iron with coal, we must have more iron in the charges, in order to have the right percentage of coal to spread over all the iron, and to be strong enough to distinctly separate the charges of iron. For a small 30″ cupola, as shown, when coke is used, the charges of iron may be five hundred pounds each; but where coal is used the charges could be twelve hundred pounds of iron.*

As to which requires the largest percentage of coke or coal to melt iron, there seems to be a great difference in practice and in opinions, but in many cases the quantity of the fuel is the regulator. As a general thing with melters that weigh their fuel, and sometimes change from one fuel to another, they use the same weight of coal as they do of coke. In melting a heat of four tons in the cupola shown, there should be twelve hundred pounds of coke used to do it, or with coal the same weight of twelve hundred pounds would be used. Again, there are places where they will use a less percentage of coal than of coke, but as a general thing the percentage is a trifle larger of coal, since it takes a little more to make the bed the height required. The reason for not here illustrating a cupola of larger diameter is because the writer believes there is more skill required to successfully melt iron in small cupolas than in a large one, and therefore the small cupola is best to here treat of. For a large cupola will stand some improper handling, and show no very bad results, but any improper management in a small cupola will be sure to cause more or less trouble.

The weights of the charges in a cupola will permit much variation without bad results. Among half a dozen cupolas

* In years gone by melters would mix the iron and fuel together when charging a cupola, but such a practice is not followed by any at the present time.

like the one shown, it would not be strange to see them charged with different weights of charges. In melting iron, the beginner must observe the following rules. If all coke is used, be sure it is about 18″ above the top of the tuyeres, and burning evenly throughout. Upon this bed put the first charge of iron, which can be from five to fifteen hundred pounds; then, if there are any heavy pieces of iron to be melted, put them in the second charge, since if heavy pieces of iron are placed in the first charge, or upon the bed, there is danger they will sink to the level of the tuyeres, and from this cause a cupola will soon get choked; the weight of the second and of remainder of the charges should run from five up to ten hundred; and to be safe as to the amount of fuel, use fourteen pounds of coke to every hundred pounds of iron, if the charge is ten hundred of iron, and let one hundred and forty pounds of coke be used between them. If all coal be used, let its bed be about 12″ above the tuyeres, and let the first charge range from fifteen to twenty-five hundred pounds, and the remainder of the charges range from ten hundred up to twenty hundred, the percentage of coal between the charges being about the same as that of coke. After two or three heats have run off, commence to use less fuel, and at the same time carefully change the weight of charges until the best results as regards economy of fuel and hot iron are obtained. If the cupola is a larger one than shown, have the fuel the same height above the tuyeres, and use the same percentage of fuel between the charges, and also grade the charges of iron heavier in proportion as the cupola increases in diameter. Some make the charges the same weight through the heat, while others will make every other charge lighter. For the last charge or two of iron the percentage of fuel can be decreased, that is, if the bed is in no danger of getting down too near the tuyeres before the last charges are melted. There can be more economy in fuel practiced by having heavy charges than by light ones; but it is not every foundry that can work heavy charges, on account of their having different grades of iron to melt during the same heat.

In melting iron it is often a great benefit to use a flux so as to clean or separate the impurities from the iron, and at the same time make it more fluid. A great many foundries use limestone or oyster shells as a flux, while others will use fluor-spar. There are also some patent fluxes in the market, for which great merit is claimed. Among them is one patented by *Edward Kirk, of Oswego, N. Y.*, the author of an instructive work on *Founding of Metals.*

The patentees of fluxes claim that the use of oyster shells or limestone destroys rather than benefits the iron, and their fluxes do the iron good by making it stronger and softer, while the use of too much limestone or oyster shells will make the iron hard; yet it answers the purpose intended in some cases, but oyster shells are better than limestone. In fluxing with either limestone or the oyster shells, they should seldom be used until the latter part of a heat, as they will then help to clean out the cupola and make it drop better. Limestone should be broken up to egg size, and thrown among the iron; a riddleful being sufficient to flux a heat of three or four tons. A shovelful of oyster shells thrown in on the last charge of fuel is a good thing to help to clean out a cupola, as it will glaze the lining and make the cinders easier to pick out. FLUOR-SPAR or any of the above fluxes are not essential to be used in the fore-part of heats unless poor fuel or dirty iron is used; see p. 310, Vol. II. The melting point in a cupola is that portion of the lining which is burned out more than the rest, and also that point at which from the highest temperature there is the most melting done. The melting point or portion of a cupola ranges from 6 inches to 2 to 3 feet above the tuyeres, and the iron is sometimes melted in the lowest portion, as well as at the middle or highest point. As the height at which iron may melt is often sufficient to contain two or three small charges of iron and fuel, it is not a hard matter to see the cause of different grades of iron getting mixed.

There will be more damage done at this point by blowing, or having the blast on, when the iron is all melted, than a dozen heats would do if the bottom is dropped, and having, according to the size of the cupola, from two up to ten hundred pounds of iron in the cupola.

Wherever the blast goes in a cupola, it cools off the fuel, and the melting iron, dropping down from above, falls upon this lifeless fuel and is soon chilled by the direct force of the cold blast. This state of affairs, from the beginning to the end, keeps all the time getting worse in the large cupola as well as in the small one. In the large cupola, however, there is more chance of the blast losing its force and coldness before it reaches the center, so that the fuel is given a better chance for thorough combustion. This permits of running a large cupola longer than a small one can be run.

As soon as a cupola begins to get black and cold at the tuyeres, we say it is choking. This is true ; and not only is the entrance being choked, but in the course of time the whole surface parallel with the tuyeres will be in the same condition, and as it increases the slower will be the melting, until no melting can be done. We then drop the bottom and have a good time trying to get a hole through the choked cupola.

There are other things besides blast that help to choke a cupola, such as improper charging, dirty iron and fuel, etc.; but, allowing that everything is done right, the cold blast will of itself accomplish it in the course of time.*

* This refers to cupolas that are not "slagged out." For information on how to run long heats, see p. 310, Vol. II.

Whenever a tuyere is getting dark or choked, it may be opened, the cold black fuel and chilled iron be driven with a bar towards the center of the hot fuel; this may reheat the cold fuel and iron. It is occasionally a good plan to stop one of the tuyeres at a time with valves, etc. This will prevent the cold blast from getting in at this point, and allow the fuel to become partly rekindled, after which the tuyere is reopened, and another one stopped up, and going thus all around to every tuyere.* For very large heats this operation might require to be repeated several times. In charging a cupola it should be kept full of iron until it is all in, and at the latter end, should there be any serious signs of its becoming choked, it is the best plan to drop the bottom if possible. In order to have cupolas work well, cleanness in their management is of the utmost importance. The fuel should be free from all dust or dirt, and the iron have as little sand on it as possible.† To look on a cupola staging is proof of the working of a cupola; if everything appears in a dirty and disorderly state, it is in most cases safe to conclude that the melting is not done in a scientific manner. Whatever knowledge there is on the subject of melting iron has for the most part been obtained from individual investigation and practice; and that man is the best melter who has studied the *cause* and *effect*, and reached a careful and well-founded conclusion in everything that has to be done from the time he begins to pick out his cupola until he drops the bottom.

* This plan is one that involves considerable stopping of the blast, and with many forms of tuyeres is not practical. For more advanced ideas in regard to keeping tuyeres open in running long heats, see p. 310, Vol. II.

† If there is much dirt or scale charged with the stock, it will cause a cupola to bung up readily, and to prevent this a flux must be used and the cupola slagged out.

TAPPING OUT AND STOPPING UP CUPOLAS.

THERE is nothing that will at times cause more excitement in a foundry than the tapping out and stopping up of the cupola, and sometimes the situation is more serious than comical. The comical part is to see the melter, when the cupola is nearly or quite full of iron, tapping out into two or three small ladles, and when he goes to stop up, the clay falls off the bod-stick or gets washed away. The iron flows over the ladle, and a spark finds lodging down the man's back that is holding it, and he lets go the ladle to dispossess the hot lodger. The foreman, who is standing by a large casting being poured, yells out for the cupola to be stopped up; the melter gets excited, runs the bod-stick without any clay on it into the running iron; the sparks fly and the iron runs around his feet. He thinks of his home or family, and gets out of it as soon as he can. The foreman, thinking the situation is getting serious, runs from the riser he is watching to go to the cupola, and when half way there he hears yells for water and sand, and, looking back, he sees the cope strained and iron running out; at which point, if he is a man that swears, he will exhaust the whole vocabulary in a very short time. His orders to stop the blast; get water go to the fire alarm—are no sooner issued to some trembling being than he hears the moulder cry out for more iron, and looks towards the cupola, at which moment a sight of his face when he sees the cupola empty and standing in a pool of boiling iron, would never be forgotten.

Such occurrences as these are frequent. I have seen

men burned, castings lost, and the shop in great danger of being burned through excitement around a cupola. Sometimes it will be caused by the iron not being carried away fast enough, but in most cases it is the melter's fault. Go into some foundries and you will see the melter running his tapping bar into the tapping hole, as shown (in cut) at *D*. A stranger seeing him would think that he was trying to knock or push in the front breast.

The position of the tapping bar as shown at *X*. is, I think, a more scientific one, for instead of trying to ram the clay into the cupola, the bar should be held so as to dig it out, or tear it away at the outside edges of the hole, so that the pressure of the melted iron will push out the center. In tapping out this way you are always digging out the old stopping, and keeping a clean hole, and doing it with less labor, sledge hammering, and burning away the tapping bars, than in any other way I know of.

I have seen melters have their tapping holes, before a heat was through, choked so badly that every time they tapped out they would have a man or two striking or knocking the bar into the breast with heavy sledge hammers, and when in it would take four or five men to pull it out.

Of course, there is sometimes iron and scrap used that will make a deal of slag, and it is hard work to keep a breast or tapping hole clean, and melters are often exhausted in trying to do so.* If they would only once adopt the plan here described, they would be astonished at the ease with which they could do their tapping.

In tapping out a cupola for the first ladle, there is often trouble on account of the iron not melting as fast as it ought to, or as fast as it will after a few minutes. The iron, especially if hard, chills in the hole, and when tapping out I have often seen the whole breast knocked in to get the iron out. To remedy this, take a one inch round core, 4″ or 5″

* This has reference to cupolas that are not slagged out through a regular slag hole in the rear of the cupola.

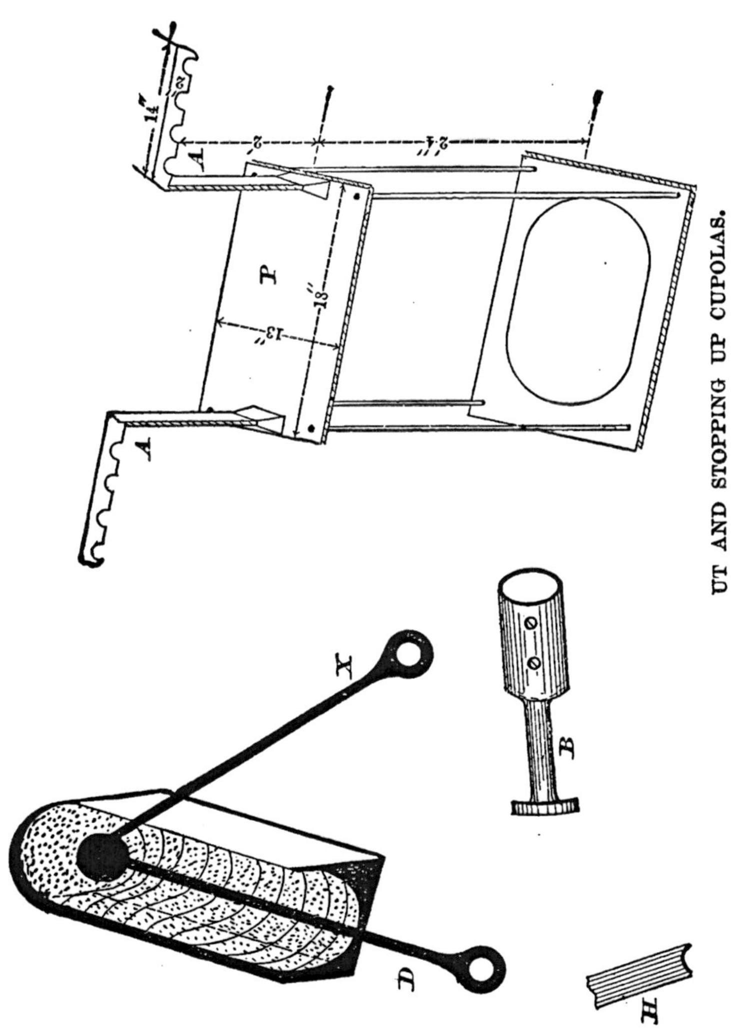

UT AND STOPPING UP CUPOLAS.

long, made with plenty of sea-coal or blacking, and when the iron is melting or at a fair stream, take the core and push it into the hole, stop over it, and when you take out for the first ladle you will have no trouble.

There is another thing a great many melters have a habit of doing when stopping up a cupola. That is, they will push the stopping against the running stream to get the hole stopped up, which always causes a splatter, and sometimes washes the stopping off the stick; whereas, if they would hold the stopping above the stream, and when near the hole push it down on a slant, they would not be so liable to burn any one, let the ladles flow over, or, worse yet, let all the iron run out on the floor, which often results in large loss.

The mixtures of stopping have often a deal to do with accidents and trouble. About the best stopping for ordinary purposes is new moulding sand dampened with clay wash. This will not make the iron fly, and will tap out easy. If clay is used, it is a good thing to mix in some horse manure or sea-coal. This will keep the clay from baking so hard, and make it tap easy.

Instead of having the bod-sticks all wood, the cut *B* shows an iron nipple made to fit on the end of a wooden stick. This will save sticks, and should the stopping fall or wash off the iron it will not fly so much. At *H* is shown a gouge-shaped end of a steel tapping bar, which is very handy.

The stand shown is a rigging that I made one day in open sand after the blast went on. The top plate was cast first, and four half inch round rods cast in it, and when set enough turned upside down and the rods cast into the bottom plate. In the top there is a pocket, cast for holding the wrought iron arms *A, A*, which were made of ¼″ iron, with a shoulder to keep them from dropping down. When not in use, they could be reversed or taken out of the way.

On the top, *P*, is a box for holding the stopping, and underneath is kept a pail of water for dipping the bars into. The arms, *A*, *A*, form a rest for holding the tapping bars and stopping sticks. The stand complete does not weigh over seventy-five pounds, and I find it handier than using barrels, boxes, or things commonly used for such purposes.

Referring back to the question of trouble with tap holes, a good plan to follow whenever practical, is not to stop up at the first running of hot iron, but to let it continue running as fast as it melts until the tap hole has been thoroughly dried throughout its body by the heat of the running stream. The metal as it runs can be caught in a bull or crane ladle, having coke or charcoal dust in it to keep the metal hot. A stream of metal running after this plan for ten to twenty minutes, will dry out the body around a tap hole pretty thoroughly, when the stream can be stopped to give the best assurance for easy tapping and no trouble with a tap hole for the balance of the heat. Sometimes melters experience trouble with slag closing up a tap hole at the beginning of a heat. This is often caused by the material used for forming the tap hole being of an easily fusible character, and which often comes from using common clays or moulding sands to form the tap hole or front breast of the cupola. By substituting a fire clay mixed with sharp sand of a consistency which would stand well for daubing the melting point of a cupola much of such trouble can be avoided. Some make their tap holes in a core box, rammed with mixtures of clays and moulding sand and dry them in an oven, and then, when making up the breasts, place a core in the breast and ram a sand and clay mixture around them. This is often a good plan to follow where trouble is being experienced with tap holes and one that can annul the necessity of letting a stream of metal run 10 to 20 minutes at the beginning of a heat to insure a good tap hole as mentioned above.

AIR FURNACES.

To many foundrymen the air furnace is a stranger. There are very few shops that have them. They are used for melting heavy bodies of scrap iron, and for melting iron for heavy castings. The difference between melting iron in air furnaces and cupola is, that in the cupola the iron is melted by being mixed with or on top of the fuel. To have sufficient draft to cause a high temperature the air is forced into the fuel by the aid of fans or blowers, but with melting iron in air furnaces the fuel is entirely distinct and away from the iron ; and to get a sufficient supply of air to combine with the fuel a very high chimney is used, the fire being at one end and the chimney at the other. The iron is melted by having the flame and heat drawn over it. The fuel that will produce the most flame is the best, and there are different styles of furnaces in use. Some have the iron piled up at the end nearest to the fire, and the iron as it melts runs down an inclined bed into a well from which it is tapped into ladles. Another style is to have the iron melted at the end furthest from the fire, and as it melts it runs into a basin midway between the iron and the fire. The author worked in a shop that melted up old brass in a small furnace after this manner. Although furnaces differ in construction they all do their melting by having the flame and heat drawn in among the iron. The style of furnace section shown in the cut is similar to that in general use. A furnace of the dimensions given should be capable of melting from twelve to fifteen tons of iron. The charging door, chimney, and firing places are very seldom situated alike, since the laying out of the shop and surroundings will cause changes. Sometimes a furnace can be easier charged if the door is at the end instead of the side, as shown, and for facility for firing it is sometimes better to have the door on the end. The firing door is sometimes on the end, and the charging door on the side, and again this order will be reversed. When the charging door is on the end, the chimney is then on the side, and there is more economy in fuel to have the chimney on the end, as more space can be used to hold iron. Chimneys should have about the same area inside as the grate surface contains, and should be high enough to give a strong draught. Some furnaces will require a higher chimney than others, on account of the shop being located in some valley, or alongside of some hill or bank. Charging doors should be made so that they can swing to and from the furnace, and be as large as possible. The author has seen a furnace that used for a door one side of the furnace, which opened for over half of its bed's length, and after the furnace was charged the opening was all tightly built up with fire-brick. A very important feature in constructing furnaces is to have a good solid foundation under them. On one occasion when the melter went to tap out he was surprised at finding there was no iron in the furnace. It had found a weak spot, and suddenly it leaked out. It proved on inquiry that this same thing had happened once before. It was stated to the author that there must be over thirty tons of iron buried below the furnace, all caused by its having a poor foundation. It depends upon the nature of the earth how deep down the foundation should go. The stone or brick foundation is built up within about one foot of the top bed, so as to allow a good bed of sand to be made to form the bottom with, as shown. Some will lay in the middle of the stone or brick foundation a coke or cinder bed, to help take the gases and steam off from the bottom sand. In building a furnace the best of fire-brick should be used for forming the inside with, and furnaces should be made with at least a twelve-inch wall; while the inside eight inches must be fire-brick, the outside four inches could be built up with common red bricks, and the top of the furnaces should be built in the form of an arch, and the whole furnace should be well bound with cast-iron plates and binders bolted together. Care should also be taken that no opening or crack exists in any form, since if any cold air gets into a furnace while it is in heat it will be apt to make trouble.

When preparing a furnace any parts that may be burnt out are daubed with fire-clay. There is a melting point in a furnace as in a cupola, but in the furnace it is that point which is on a level with the iron when melted. It does not burn out so much as cupolas do, but nevertheless it requires as careful daubing as a cupola does. In putting a bed or bottom in a furnace, it can be raised or lowered according to the amount of iron required to be melted, varying from two to five tons more than its average. The lowest point of the bed is at the tapping hole, and the highest point is at the chimney entrance, ranging from 6″ up to 12″ higher than at the tapping hole. The sand used for making the bed is similar to that used for cupola bottoms, but if anything it should be more open. At the highest portion of the bed, where it runs in under the chimney, the sand should be a little closer or loamy, for when the sand is of a sharp nature the current of the draft is sometimes strong enough to wash it with it. In mixing this sand care must be taken not to have it any damper than sand used to mould with, and in forming and ramming the bed it must not be rammed too hard. After a bed is evenly formed and given the right slope, inch boards are then laid over the top of the sand bed so as to protect it from being cut up with the iron. When the furnace is

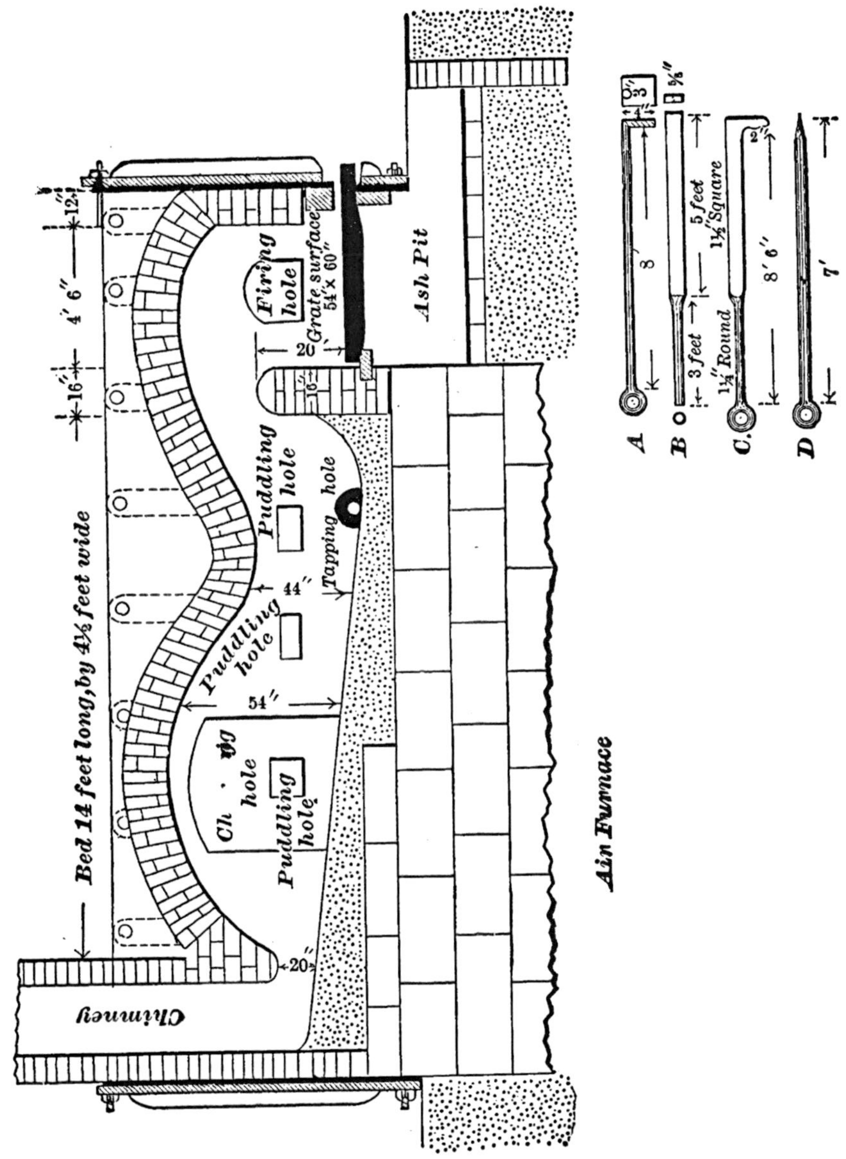
Bed 14 feet long, by 4½ feet wide
Chimney
Firing hole
Grate surface 54″ x 60″
Ash Pit
Puddling hole
Puddling hole
Puddling hole
Tapping hole
12″
4′ 6″
16″
20″
44″
54″
20″
A
B
C
D
8′
5 feet 1½″ Square
3 feet 1¼″ Round
8′ 6″
7′
Air Furnace

being charged up, unless dirty iron is used or a furnace does not work well, a bottom will often stand for two or three heats.

Making the tapping hole, breast, and spout, must be done in a reliable manner, and the size to make the tapping hole depends upon the class of work to be poured. If the iron is to be tapped out into crane ladles, the hole should not be any larger than 2″ in diameter, but if the iron is to be tapped out in a large basin that will hold all the iron there is in the furnace, then the tapping hole should be about three inches in diameter. A plan for stopping up the hole with sharp sand, so that it will tap out without any danger of bursting in the breast, is shown in the article entitled "Reservoirs and Ladles for Pouring Heavy Castings." When charging a furnace, the iron should be kept back a foot or two from the tapping hole, for when it is close to it there is apt to be serious trouble, since the first iron that melts runs to this point because it is the lowest; and if there is iron there the metal is apt to become chilled.

Light scrap, pig iron, or any iron that melts easily should be the bottom or first charged iron, and heavy rolls or larger lumps of scrap should be the uppermost or top iron, as the upper iron gets the most heat, and thus we have the iron that is the easiest to melt on the bottom, and the hardest on the top. The light and heavy iron will melt proportionately, which is one of the main things to accomplish in melting iron in a furnace; for if when the iron is most all down, there are found one or two pieces that are not down, more fires will have to be made in order to get them in a fluid or melted state, and these extra fires are apt to harden the iron, or burn the life out of it. Iron should also be charged as open as possible, so that the flame and heat can get at the greatest amount of surface.

What is here meant by the iron being down is that when

looking into a furnace by removing the loose bricks from the puddling holes, we show that there are no lumps of iron to be seen. Connected with a furnace there are the tools *A*, *B*, *C*, *D* shown. Whenever there appears to be any lumps of iron that are not melting as fast as they should do, in order to be down with the rest the melter will take the poking-down bar *D* (of which there should be two sizes), and break the half-molten lump into as many small pieces as possible, and then with the puddling and pulling-down bar *C* he will move the lumps into the deeper and lower metal, and there work it round for a while. This work should be done quickly, as the leaving open of the puddling holes allows cold air to get into the furnace. When the iron is all melted and about hot enough, which is ascertained by dipping some out of the furnace with a small hand ladle, the melter then takes a long stick of wood, and, putting it through the different puddling holes in turn, he mixes or polls up the iron from five to ten minutes, after which, if the iron is hot enough and everything ready, the furnace can be tapped out.

For fuel, bituminous coal is the best, as it makes much flame. Anthracite coal or coke may be used, but not with as good results as with the soft coal. With anthracite or coke there would have to be some blast used. In starting a fire, try to have a little good, clear coal upon the grates for the first two or three fires, as it will help to keep the clinkers from forming on the grates, after which a poorer grade of coal might be used. In firing up there should be some system, so as to keep up an even fire. About every fifteen minutes the fire could be supplied with fuel, and about every five minutes before firing take the bar *B*, and by running it in between the grate bars loosen up the coal. *A* is a bar for leveling the coal over the fire grates and raking up the same. The grate bars are all made single, sometimes being wrought iron as well as cast. As a general thing, if the furnace has not worked well, or not been managed rightly so as to have the iron on the chill side when it is all melted, it is a hard matter then to make it hot; and to have hot iron it must be melted hot as it comes down. Trying to make dull iron hot, after it is melted, is nearly like trying to make cold water boil by having the heat pass over the top of it. Air furnaces are good for producing good, strong iron when properly managed, but there are very few cupola melters that would be able to successfully run an air furnace. There have been some very bad blunders made in handling furnaces; often they have had to be torn almost to pieces in order to get solid frozen masses of iron out, which came from improper management during the melting. To be a good air furnace melter a man must have brains and practice; and unless a firm has large, heavy iron that they wish to make a business of melting, it is better to erect one or two large cupolas, for they will not only melt iron with a much less percentage of fuel, but also avoid risks which have often to be taken in melting iron in air furnaces.

BLACKING MIXTURES.

It is of great value to a moulder to have a mixture of blacking that will peel and make a smooth skin of a dark blue color on a casting, and the failure to get it is not always because of improper blacking, but due to the method of mixing. Clay wash, molasses water, and sour beer are liquids that are generally used to mix up blacking, and their proportions can always be regulated so as to control, in a great measure, the quality of the mixture. In mixing blacking for thin castings the clay wash or molasses water should not be so strong as in the case of castings over one inch in thickness. Too much clay in any form in blacking is a bad thing, as it closes up the pores of the blacking, and is very liable to scab. Molasses water is valuable to mix blacking with, but care must be used as to the quantity, as too much molasses will cause it to flake and crack when the heat of the metal reaches it, and when the casting comes out, if it is a heavy body, it is apt to look veined and streaked, as a heavy green sand casting appears when the facing sand has been too strong. A half pint of black molasses is as much as should be used to a common-sized water-pail of mixed blacking; anything in excess of this amount is apt to cause trouble of some kind. Sour beer is also of use in mixing blacking, and will cause trouble if too much clay is mixed with it. Castings will often peel better by using blacking mixed with all pure water than if clay was mixed heavy with the water. It is, however, necessary to have some clay in the blacking to peel properly heavy castings, but before mixing any we should closely examine the dry blacking to see how much it contains, as there is generally more or less mixed in with blackings when originally made. Some blacking contains so much clay that it does not require any clay wash added in mixing it for use. The finer blacking is ground, the better mixture it will make, and good blacking when mixed will not settle down to the bottom of the pail, but will grow thicker in time. If the blacking is too light it will float on the top while mixing it, and such a blacking should be seldom used; on the other hand, a blacking that is so heavy as to sink to the bottom when mixed, generally contains much dirt or clay, and should also be rejected. If you wish to make a nice, clean, skin-colored casting, below are a few recipes for mixtures which have been proved satisfactory:

1 of Lehigh blacking,
$\frac{1}{4}$ of charcoal blacking,
$\frac{1}{4}$ of German or American lead.

Wet with beer or molasses water, slightly colored with fire-clay. This made a good mixture for cylinders and engine castings not over 3″ thick.

$\frac{1}{3}$ pail of heavy prepared blacking,
$\frac{1}{4}$ pail of lighter prepared blacking,
2 handfuls of flour,
1 handful of salt.

Wet with beer colored with fire-clay.

Molasses was also used, but beer worked the best in this mixture. The salt was put in to harden it, and make the blacking dry rapidly, and the flour to give it a body. The salt part of this mixture was not altogether satisfactory, as it somewhat prevented a fine finish, otherwise the mixture was entirely satisfactory

$\frac{1}{2}$ of heavy prepared blacking,
$\frac{1}{4}$ of light prepared blacking,
$\frac{1}{2}$ pint of good clear oil.

Wet with beer colored with common clay. The oil was put in to harden the blacking, and also to cause a smooth and easy finish, and this mixture made a nice-colored skin on locomotive cylinders.

To mix a blacking that will peel a heavy solid casting, such as anvil blocks, rolls, or heavy cannons, it is a good plan to take either a pure Lehigh or a coke blacking and mix it with one third of *plumbago*, or, as commonly called, *black lead*, and wet the mixture with black molasses water colored with fire-clay ; then, after the face of the mould has been roughly sleeked over once with the tools, take some plumbago and wet it with molasses water.

Make the mixture thin, and go over the mould with the plumbago blacking by using a camel's-hair brush. Next dust from a bag or spread on by the hand a light dust of dry plumbago over the mould, and after this finish up the mould. By this mixture and plan heavy rolls and anvil block castings will drop the loam or dry sand without touching them, and the skin and color will be beautiful and perfect if properly done. All blacking contains more or less carbon, and the larger the percentage the more heat it will stand.

Any substance put on the face of a mould which will prevent the hot liquid iron from burning or eating into the sand and not SCAB, will help to make a smooth skin or surface on a casting.

Plumbago blacking contains more carbon than any other in use, and it is said the highest temperature will not melt, soften, or change its condition ; therefore when used on the surface of a mould these results are obtained. All blacking is improved by being mixed a day or two before required, and in mixing blacking it should be screened from one pail into another several times, that the different parts may be thoroughly mixed and clean.

LOAM MIXTURES.

THERE are certain sands which can be obtained in almost any section of the country, and from which, if used according to their clayey qualities or sharpness, mixtures of loam can be made. There are two classes of sand which generally combine in order to make a loam; one is of a close, clayey nature, and the other a sharp or coarse open-grained sand. The clayey sand gives a body to the loam, while the sharp sand makes the loam open and porous, so that the iron will kindly lay against its surface.

This subject is more fully treated in the chapter on "The Surface of Loam Moulds." The following are a few loam mixtures which have worked well, and are given to show the proper proportions of parts, and the method of mixing loams:

3 pails of fire sand,
2 pails of moulding sand,
1 to 10 of horse manure.
Wet with thick clay wash.

4 of fire sand,
1 of moulding sand,
1 of dry sieved fire-clay,
1 of white pine sawdust.
Wet with thin clay wash.

For a finishing loam, the same mixture would sometimes

be used; the only difference is, that the clay and manure should be left out, and instead of putting the sand through a No. 4 riddle, it would be screened through a No. 8 sieve; and again, 1 part of fire sand, and 3 parts of moulding sand would be used, and the mixture wet with beer. If, however, the moulding sand was not too close, it could be used by itself if wet with beer.

Mixtures of loam containing fire-sand are in general used only in the Eastern and Middle States. The following mixtures are of a Western origin, although similar mixtures are often used in the East:

4 or 5 of loam sand, according to clayeyness,
1 of lake sand,
1 of manure.
Wet with medium clay wash.

Finishing loam is the same, only screened through a No. 8 riddle.

Mixture of loam used for making thin pulley patterns:

2 of fair loam sand,
1 of old burnt loam,
2 of lake sand,
1 of manure.
Wet with very thin clay wash.

In the first casting of these pulley patterns only the ordinary mixtures were used; but when the moulds were cast, the iron blew so hard that but little was left in the mould. With the above weak mixtures, however, the castings came out all right. It is sometimes a good thing to use about *one* to *twenty* of sea-coal in loam for light thin castings. The following mixture

1 of strong loam sand,
1 of coarse lake sand,
1 to 6 of manure,

wet with water, proved very bad, because the loam sand was so clayey that it took too large a quantity of coarse lake sand to make it open enough to use. In any loam mixture it is not well to have to combine sands which are very close and very open, or have to mix coarse sands together in order to make a mixture that will work satisfactorily. The nearer to an even nature we can get the sharp and the clayey sands, when the two are mixed together, the closer we approach a natural loam.

The great difficulty in using finishing loam mixtures is, that they generally close up the pores of the under or coarse loam too much, and thereby render a mould liable to scab. The following mixture gave results very satisfactorily for heavy castings, as the casting came out as smooth as a piece of thin stove plate, and this same mixture was used for the finishing loam on swept-up rolls. It was mixed as follows:

2 of old dry sand,*
1 of strong loam sand,
1 of lake sand,
½ coke dust or sea-coal.
Wet with water.

In connection with the following loam mixture the name

* This was taken out of a dry sand mould mixture, having been used once, the life was partly burnt out of it.

of Mr. William Fitzsimmons must be mentioned, since he has accomplished by his own genius many valuable results in foundry practice. His loam mixture, which works well, is one that can be made in most any section of foundry practice.

$5\frac{1}{2}$ of lake or bank sand,
$2\frac{1}{2}$ of moulding sand,
$1\frac{1}{2}$ of horse manure,
3 of clay wash.

This loam is mostly sharp sand, and to give it strength the clay wash is used. This clay wash is mixed in such a manner that there is a certain quality of clay in every batch without fail. The mixing of this clay wash is the most important part of the mixture, and must be measured very exactly by the following plan: Take a large barrel that will not leak, and for every well-packed pail of common clay put in two full pails of water, and then for every three pails of clay put in half a shovel of flour; this will help to thicken and ferment the clay wash. Then the whole should be allowed to stand over night, so as to soak the clay soft. In the morning all the clay is thoroughly mixed. When the sand is all ready to be wet the three pails of clay wash are taken from the barrel and mixed in it. Should the sand be unusually dry, so that the three pails of clay wash would not wet it enough, use for the balance water. If the sand be very wet, use a stronger proportion of clay in the wash. If required to use this loam after it is old, always wet it with water. This loam, and, in fact, any loam is better if mixed two or three days before using, for it is tougher and more of a loamy nature.

If a stronger loam is desired, only use seven pails of sand to three pails of clay wash. For a finishing loam mix the same proportions of sand, but instead of the horse manure use cow manure, for it makes a smoother-skinned casting. Take 1½ pails of fresh manure, and mix the three pails of clay wash with it, rubbing the manure and clay through a No. 4 riddle; then mix it with sand which has been screened through a No. 8 riddle. The horse manure can be used in place of the cow manure, if more accessible.

The following is a mixture of loam which can be made from moulding and lake or bank sands ·

1 of moulding sand,
1¼ of bank sand,
1 to 20 of dried sieved fire clay,
1 to 6 of horse manure.

Sometimes one to twenty of coke dust or sea-coal is mixed with the loam.

This loam was wet with good clay wash, and worked well on the castings.

An odd kind of sand is sometimes found, resembling meal, and looks very much like fire sand, except it is not so red. It is more loamy, however, and has nearly as much body as moulding sand. Mixed as follows it made a splendid loam:

5 of the meal sand,
2 of lake sand,
1½ of horse manure.
Wet with medium clay wash.

All the above clay washes (except when fire-clay is named) are made from common red clay, and what is here meant by loam sand is a sand which contains more clay in it than moulding sand, making it of a loamy nature.

The lake or bank sand comes under the head of sharp sand, and is always used for an opener.

There are two bad features mixtures of loam sometimes possess: one is, it will not stand the dropping or washing of iron on it; and the other, it will scab.

Iron borings or filings are useful to use sometimes in loam to keep it from cutting, or a little flour will answer; of course this should only be used on that part where the iron strikes it directly, as if used in any other part it might render it liable to scab; a mixture of loam can be so made as to both cut and scab by making it of a close, weak mixture. Very fine-grained sands will generally scab and cut easier than open-grained sands. If a mixture of loam can be made from open sand having body enough to stand the iron, it is better than to use a close-grained sand in order to give it strength. The best mixed loams are those which will stand the dropping of iron and not scab in any part of the mould, and to obtain such a mixture depends much upon the mixer's judgment and the quality of the sands which he has to use.

DRY SAND MIXTURES.

DRY sand moulding is in many respects similar to loam moulding.

A cut or a scab on a dry sand casting is the result of similar causes as scabs or cuts on loam castings, and the mixture of loam used generally calls for about the same proportion of sand for making dry sand facings. What is meant by facing sand is, that sand which forms the surface of a mould, and its thickness ranges from one to two inches; the sand which is at the back of this has not such care taken with its proportion or mixing. This backing sand answers very much the same purpose as the bricks in a loam mould, giving support to the surface. Backing sand should be worked as open as possible, so as to allow gases of the surface sand to escape through it as freely as possible, and the facing sand should be worked as open as its strength will permit. The dampness of the sand should not greatly exceed that of green sand, as the wetter the sand when used, the harder and closer will it be when dried. A dry sand facing frequently is made wetter for some jobs than for others. If a mould takes three or four days to ram, or if that time elapses before it can be finished ready for the oven, the mixture should be made damper than if it was to go in the oven the same day it was mixed. In making dry sand facings it is better to have them well tramped and mixed, as by so doing it will give strength and toughness to them. The following are a few mixtures that will give the proportion and afford an idea how to make dry sand mixtures:

12 pails of lake sand,
12 pails of strong loam sand,
4 pails of moulding sand,
1 to 10 of coke dust,
$1\frac{1}{4}$ of flour.
Wet with water.

The above was used for the teeth of a large spur-gear wheel, and it worked well. This mixture would be too close for flat surface moulds, but for the teeth of gear wheel sand, it is better if worked closer, for the teeth will hang and mould up better, and as long as they are well vented there is very little risk of scabbing.

The following mixture was used for making a large bevel wheel, and the sand was made more open because the teeth were on a bed, and therefore there was no danger of their dropping. The sand could be made more open, and thus lessen the danger of scabbing, which is a thing dry sand bevel-gear wheels are sometimes liable to do.

The Jersey sand here mentioned is similar to a fine lake sand, except it is whiter; it is a sand somewhat like fire sand, and has more of a body to it than lake sand.

1 of moulding sand,
1 of Jersey sand,
1 of fire sand,
1 to 16 of sea-coal.
Wet with thin clay wash.

The following mixture made a close facing, but was one where there was danger of scabbing:

1 of close loam,
1 of open loam,
1 to 16 of sea-coal.
Wet with clay wash.

There are sometimes places where a loam sand cannot be procured; in such places the mixture below will be found to give satisfaction.

6 pails of moulding sand,
1½ pails of lake or bank sand
1 to 30 of flour.
Wet with clay wash.

This same mixture was used for backing, only it was not mixed so carefully, and about 1 to 40 of flour used. With this mixture cylinders, as well as jobbing work, have been cast.

Another mixture for cylinders is :

4 of fair loam,
1 of lake,
1 to 14 of sea-coal or coke dust.

Wet with clay wash according to clayeyness of the loam, in fact in any mixture where clay wash is used its thickness should be regulated by the nature of the sand. The backing used with this facing was 5 parts of loam and 1 part of lake wet with clay wash.

A mixture that can be made most anywhere, and is good for ordinary work, is as follows :

1 of moulding sand,
1 of bank sand.

Use 1½ of bank sand when it is wanted open, and 1 to 30 of flour, 1 to 20 of blacking, wet with clay wash, and for the backing the same proportion of sand was used having about 1 to 30 of flour, omitting the sea-coal blacking.

The following mixture was made because a very clayey

loam had to be used, and to make it work there had to be a larger proportion of coarse lake sand.

6 of strong loam sand,
6 of lake sand,
2 of old dry sand,
1 to 40 of flour,
1 to 14 of sea coal.
Wet with water.

The backing was mixed from half lake and half loam sands; the whole mixture was then used for cylinder casting.

In the same shop rolls were made by being swept up. A good mixture for the grooves was as follows:

2 pails of old dry sand,
1 pail of lake sand,
1 to 12 of sea-coal,
1 to 18 of flour.

Made as wet as could be worked with thick clay wash.

For the body of the rolls the old sand was used, and it was renewed as follows:

16 pails of the old sand,
8 pails of lake sand,
4 pails of new loam.
Wet with water.

This is a good proportion to use in renewing any old dry sand, as there is twice as much open sand used as loam or clayey sand. The great trouble with old dry sand is its closeness, since every time it is used it becomes more fine and dusty, hence a good thing to often do with old dry sand is to shovel it into a No. 8 sieve, and by shaking it a little the dusty or very fine portion of the sand will separate from the better and coarser qualities of the sand, which when thrown in a pile by itself and the fine dust screened out, and some new sand added to renew it, will be found to work well. All new dry sand mixtures should be mixed in proportions according to the nature of the sand and the moulder's judgment of what is required for his special job.

CORE SAND MIXTURES.

CORES are generally used to form the interior portions of castings, and the least neglect or mismanagement in making them is apt to cause trouble. There is no portion of a casting that requires such care in respect to venting. The vent of the outside portion of a mould may sometimes be confined and no harm done to the casting ; but let the vent of a core nearly surrounded with iron be confined, and the result is a bad casting. There are two reasons for the confinement of gases in cores ; the first is, they may not be sufficiently vented, or it may be improperly done ; and the second, the iron is allowed to get into the vents by not having them well secured or made air-tight in their prints.*

There are three modes of venting cores. The FIRST is by using straight rods. The SECOND, by using strings, ropes, or bands of straw or hay. This class of venting is only done in crooked cores. Sometimes the strings or ropes are coated with wax or soap, and in some cases they are not pulled out until the core is dried. Another plan, sometimes practiced in venting partially crooked cores, is to use straight rods in the straight part, letting them run through the print as far as they will go into the straight portion of the core box, without danger of coming too close to the sides of the box, and when the core is dried the crooked portion is then vented by using vents filed or scratched into it, and then passing a string or rope through the straight vents into the openings made in the crooked part, filling up the balance of the filed out openings with a mixture of half stiff blacking and core sand ; then the strings or ropes are

* Two chapters which should be read in connection with this are found on pp. 101, 108, Vol. II.

pulled out, leaving a clear vent. This plan is practiced considerably in the making of cylinder ports, or S cores, as they are sometimes called. If there are any doubts as to the clearness of the vents, they can be tested by blowing tobacco smoke through them, after which the end of the vents opposite to the prints are carefully stopped up.

The THIRD MODE OF VENTING is practiced in the making of large cores. In this case the interior portion of the core is filled with coke or fine cinders, thus saving core sand and firing, as well as affording means to carry off the vent.

The mixture of core sands depends upon the class of cores to be made. For small cores finer sand should be used than for large cores. There are two articles, *flour* and *rosin*, that are used to mix with the core sand, in order to make them firm and solid when dried. Flour is most commonly used, on account of the ease with which it will mix with sand. Rosin is good for cores that are hard to vent, as the gases escape and ignite freer than when flour is used. Rosin cores also will stand the dampness of green sand moulds, and, as a general thing, leave a smoother surface or hole in a casting than flour.*

The following are mixtures of core sands in use. The common mixture of sand for large ordinary cores is

2 of lake or bank sand,
1 of moulding sand,
From 1 to 12, up to 1 to 18 of flour.

Cores that are not to be handled much can be mixed with less flour than cores which are to be filled and lifted in and out of moulds several times, in order to make them fit. For wetting the sand some shops use an all clay wash, while others will use nothing but water, and again there are a few shops that use only beer or molasses water. This makes a good strong core, but on account of its expense it is but little

* See page 362 of this book for a description of binders other than flour and rosin.

used. For such cores as are required to be very firm and solid, some foundries do not use clay wash or water, but go to the extra expense of using beer or molasses water, which in many cases is advisable. When beer or molasses water is used, a less percentage of flour is required. In mixing sand for large cores, it is sometimes advisable to mix it half lake or bank sand, and half moulding sand. Having an excess of moulding sand will cause the core to hold together, while it is green, better than if the sharp sand is used in excess, as given in the first receipt. When the core sand is mixed half and half, as above stated, it is better to have the sand wet with some beer or molasses water, so as to give the core firmness when dried. Many places mix their sand for very large cores as follows :

3 of lake or bank sand,
1 of moulding sand,
1 to 14 of flour.
Wet with clay wash.

But cores thus made should be well rodded and rammed, especially if the core is one that stands up straight. Such a mixture will stand a hot fire better than if more moulding sand is used in it.

For making hard fine small cores a good mixture is :

3 of moulding sand,
1 of lake or bank sand,
1 to 14 of flour or rosin.

Wet with molasses water, mixed about 1 to 20, or one pint of molasses to a pail of water.

In using rosin, pound it into a fine powder in an iron kettle or pot. Sometimes all moulding sand is used, when rosin is mixed in it, and again the rosin and flour are used together, the rosin being mixed with it to make the vents come off more freely.

Rye flour is often used for making core sand, and it makes a nice open core, and is also good for making paste to joint cores with, or may be used on the joints of moulds, as it is not so sticky to handle.

To show some of the different ways that core sand is mixed for special jobs, the following receipts are given:

CAR WHEEL CORES.

6 of sharp sand,
1 of moulding sand
1 to 16 of flour.
Wet with water.

INGOT CORES, FOR MAKING CASTINGS TO POUR STEEL INTO.

66 pails of coarse lake sand,
66 pails of moulding sand,
18 pails of clayey loam,
14 pails of horse manure,
2 pails of flour.
Wet with water.

This mixture made a loamy open core sand, which will hang together, with little danger of its scabbing.*

CORE SAND, FOR MAKING SEGMENT CORES FOR FORMING OR MAKING LARGE GEAR WHEELS.

2 of moulding sand,
1 of bank sand,
1 of Jersey or fire sand,
1 to 16 of blacking,
1 to 20 of flour.
Wet with thin clay wash.

* On account of this mixture being so open or weak, it would "cut" easily were metal to drop or flow direct from gates against it.

CORES IN HEAVY CASTING.

When cores run through heavy bodies of iron, the hot liquid raises the fusible element of the sand to such a high temperature that the grains fuse together, so that when the casting cleaner tries to get the core out, he finds it almost as hard as the iron. A good thing to prevent this fusing of the sand is to mix some sea-coal or blacking in it, and to give the surface of the core a good body of black lead, or plumbago blacking. This outside coat of blacking will prevent the liquid iron from eating into the surface of the core sand, and the sea-coal or blacking mixed in the sand burns away and passes off in the form of gas, leaving a porous body between the grains of sand, which assists in preventing its fusing. In putting rods in such cores as are subjected to high temperature, it is a good plan to coat them with two or three thick coats of flour paste, and dry them in an oven as it is put on ; for by doing this the dried paste burns off from the rod, and leaves it free to come out of the casting.

Of late years there has been much experimenting to find a bond for sand that would not possess the objectionable features found in flour or rosin, and which with the former lies in oven heat burning its binding qualities to make cores rotten, and then again it causes cores to swell or crack and not vent freely, while with the latter, cores remain soft when hot. At this writing, 1899, we find several core compounds on the market, of which much is claimed in having an advantage over flour or rosin. There is also on the market a binder called "glutrose," which, when mixed with flour and rosin, equal parts, or used alone, works well in many kinds of cores, especially such as are desired to crush easily to allow good freedom for contraction when a casting is cooling, as with an all flour bonded core many castings have cracked simply because the cores were so hard and would not soften with heat like glutrose bonded cores, to permit freedom for contraction.

Raw linseed oil is also being much used for a bond. This makes a strong core and one that will not readily absorb dampness from the atmosphere or a green mould. Some have improved on the use of the raw oil by mixing it equal parts with a solution by adding rosin to benzoin, all it can dissolve. Core sand wet with this bond requires baking with a hot fire, and gives a body that will vent freely, also such as not to absorb dampness, if standing long in a green mould.

Glue water is now being used by some for a bond. This makes a hard core and one that is hard when hot as well as cold; but care must be taken not to use any more of this bond than sufficient to make a core hard enough to handle well. It gives off less gas than any other known bond and vents very freely.

To use any of these late bonds for making cores requires some experimenting, for the reason that rarely, if ever, is the sand in any two shops alike in nature, and the proportion that will work well with one mixture will not do so with another.

GREEN SAND FACINGS.

THE nature of the sand with which a green sand mould is made affects the quality of the casting to a remarkable extent. To make fine light castings, finer grades of sand are used, and coarser for the large heavy castings. The main reason for using the coarser grades of sand for heavy work is, such sands generally have more body to withstand heavy heats, and again, coarser sands admit of being rammed harder with far less danger of scabbing the moulds.

It would be a hard matter to definitely show how any one could decide if a new grade of sand was suited to his special class of work, since a judge of moulding sand must be a person of some experience at moulding, or one familiar with moulding sand.

A moulder, in deciding if moulding sand will answer his purpose, generally takes some in his hand, and after giving it a squeeze, he will then hold the oblong ball by one of its ends, slightly swinging it to and fro, to test its hanging qualities, after which he will closely examine the grain of the sand by laying it on some flat surface. If he should observe too large a percentage of quicksand in it, it would not be very favorable for his purpose, especially so where the sand is to be used for making heavy castings; in fact, for any class of work too much quicksand is very objectionable. Moulders prefer a sand having a good body, and of a porous nature, for heavy work, and a fine grain sand for light work. Sometimes foundries receive sand having weeds growing in it. Such sand as this is generally rich. An occurrence that happened not long ago with sand having weeds in it may here be cited. In making some large floor plates, the first one was lost, because there were some little lumps on the cope side of the casting, and underneath these lumps were hollow places. Many reasons were given for the failure, but nothing seemed satisfactory. Another one was cast, and as the shop was only casting every other day, the mould was closed, and on the day after, just before casting time, the cope was lifted off, and then it appeared. Another bad casting would have resulted if the casting had been made, for there were little sand mounds in several places on the bottom part of the mould, and in looking under the cope there appeared small weeds growing downwards. As these weeds grew they pushed down the sand, leaving the lumps on top, and the holes in the bottom, which appeared in the first casting.*

As a general thing, sea-coal or bituminous facing is mixed in with sands for heavy casting, or for casting machinery; but sometimes coke dust is used. The mixing of these facings with sand prevents, to a certain extent, the grams of the sand from being partially melted, and prevents the hot iron from burning and penetrating into the sand. There is a limit as to the percentage of facings to be mixed with the sand, which, if exceeded on the heavy castings, causes the iron to eat into the facing sand, and leaves a casting full of sharp veins.

For light casting, too great a quantity of mixed facings is apt to prevent the castings from running sharply, or will cause it to be cold shut. Facings also have a tendency to make the skin of a casting hard. The proportions in which sea-coal or facings are mixed with sand, ranges from *one* to *six* up to *one* to *twenty*, *one* to *six* being about as strong as it will stand, so as to not have the casting look veined, and

* Another thing to be guarded against in using some new sand is worms. After the sand has been used a while, they will die by being burnt or smothered with gas, etc.

one to *twenty* is about as weak as it can be mixed, to show any effect on castings.

For light castings under three eighths of an inch thickness, facing sand is very seldom used, and for castings ranging from $\frac{3}{8}''$ up to $1\frac{1}{4}''$, there is generally one part of sea-coal or coke dust mixed in with ten of sand. From $1\frac{1}{4}''$ up to $2\frac{1}{2}''$ it is generally mixed *one* to *eight*, and all over $2\frac{1}{2}''$ is commonly mixed *one* to *seven*, or *six*. In using facing sand, it is not always the thickness of the casting which is a guide for the strength of the facing sand to be used. There are other things to be considered: the first is whether it is desired to pour the casting with hot or dull iron; the second, the distance of some parts from the point where the iron enters the mould; and the third, how long a time it takes for a mould to become filled with iron. Heavy solid lumps of castings have been known to be cold shut, from using what might be called facing weak in proportion to the thickness of the casting. Strong facing on the sides of a mould where the iron runs in and rises up slowly, will sometimes cause heavy thick castings to be cold shut. The square corners of a casting should have weaker facing sand used upon or against them than the straight plain surfaces; and the lower parts of high moulds should have a stronger facing used upon them than the upper portions, since if the strong facings were used at the most distant or upper portions of a mould, as can be done at the lower portions, or those near the gates, the castings would be sometimes liable to become cold shut. In some places, in mixing facing sand, they use one half old heap sand and one half new sand; but the majority of shops use new sand throughout. When old sand is used, less sea-coal facing is required. In mixing facing sand, it should be well mixed and riddled. A facing sand passed through a No. 8 sieve before going against the pattern, will make a smoother casting than that passed through a No. 4 riddle. A stronger facing sand can also be used on very thick castings, by having it well tramped and mixed. There are many receipts of green sand mixtures here given, placed in with the articles, under the head of Green Sand Moulding.

A mixture for one job may have to be changed for another, although it apparently looks the same, and in green sand moulding as in loam or dry sand. The moulder has often many points to consider in order to properly *make* and *use* sand mixtures. A chapter which would be instructive to be read in connection with this is found on p. 208, Vol. II.

CLEANING CASTINGS.

THE general idea of a good casting is one that looks well with the least amount of labor spent in cleaning it. Some moulders will make castings that require only half the labor to chip and clean them which others will. Sometimes gates will be cut so clumsily on small castings, that it takes longer to chip them than it does to make the mould. Or, again, the castings may be all strained and swelled, or scabbed; and when the chipper has spent more time to clean it than it took to mould it, the moulder will take the credit for its final appearance.

To properly clean castings is as essential as to properly mould them. A well-regulated foundry will always be found to have facilities for the cleaning, as well as for the moulding of its castings. If possible, castings should be cleaned in a department separated from the moulding-room; and for cleaning large castings there should be an ample supply of cape, cold sets, hand chisels, and different-shaped scrapers and wire brushes, together with a place for each class of tools, so that there will be no time lost in hunting for them.

The cleaning of small castings requires vitriol bath tubs, and tumbling or rolling barrels. The latter are generally used in shops that make small castings a specialty. The method of using vitriol, generally employed, is by means of an inclined wooden platform, having its lowest point hanging over an iron or wooden kettle, or box, and in this will be placed a mixture of about one third vitriol and two thirds water. The castings are then placed on the inclined platform, and a long-handled iron dipper is used to spread the mixture of vitriol and water on them. They are then left to dry until they appear of a whitish-looking color, the time required for this being from eight to twelve hours. If one application of the vitriol does not remove or loosen all the scale, the process must be repeated.

By the side of this inclined platform is usually placed a kettle, or oblong wooden box, with which a water or steam pipe is connected. If cold water is used after the box is partially filled up, the steam is turned on, and the water made as hot as possible, after which the castings lying on the inclined platform are placed in this hot water, and when thoroughly washed are lifted out; then, if not hot enough to dry themselves, they are placed in a box filled with sawdust, so as to prevent them from becoming rusted. Wooden boxes lined with lead are believed to be better than others, since vitriol would soon eat iron kettles. The hot water into which castings are placed should be often renewed; for to dip castings in water that has been used three or four times is apt to leave a whitish color upon some parts of the casting.

Another plan for cleaning small castings, whose form will permit, is to place them in cylindrical barrels, so constructed that the castings can be readily placed in and taken from them. In this method sometimes cinders are mixed with the castings, or the barrel may be partially filled with castings and some fine shot, and the remaining space filled with long wooden blocks. These blocks, as a general thing, are only used when there are not enough castings to fill up the barrel, or when the castings are of such a shape that the barrel could not otherwise be packed tight.

Light and heavy castings should not be put in a barrel together, as there is danger that the heavy ones would break the light ones.

Some shops which have many quite small and light castings to tumble, have a large number of star-shaped shot put in with them. These little stars are similar to $\frac{3}{8}''$ or $\frac{1}{2}''$ round shot, with four or five sharp points projecting about $\frac{1}{4}''$. The sharp points find their way into all corners of the castings, as the barrel revolves, and the castings are thoroughly cleaned by them.*

The cleaning of large castings is generally done by hand, and it is as essential casting cleaners should be neat and particular in performing their part of the work, as that the moulders should be in theirs, if a shop would have the reputation of making good, smooth castings.

Since 1897 we have what is called "sand blast," which some founders use for cleaning castings. The plan of operation is to force a stream of sharp sand through flexible pipes by means of pneumatic power against the surface of the castings. It is said that the action of the sand is very cutting and will clean castings of their sand or scale very rapidly and, in some classes of work, prove very economical. For cleaning machine or other castings of their sand or scale, that are to be painted to take on and maintain a good heavy, permanent coat of paint, the sand blast is especially efficient. Advance in founding has also given us the pneumatic chipper, a tool that can in many cases save fifty per cent hand labor in chipping light fins, gates, etc.; but for heavy lumps or gates such tools have not, as yet, been perfected.

---

* The cleaning room of the light workshop to-day, 1899, has attained systems in utilizing emery wheels, tumbling barrels, and vitriol baths for cleaning small castings that would seem all that was possible to be perfected, and in some lines have as much to do with the excellence of the castings produced as that of the sand or facings used in their manufacture.

WEIGHTS OF CASTINGS.

It is no uncommon occurrence for a moulder, in pouring castings, to have them run short of the proper amount of iron. Not of necessity from a deliberate design, but because his judgment has deceived him, either by miscalculating the amount in the ladle, or that required to fill the mould. In pouring heavy castings the moulder should seldom depend upon his judgment, for the risk is too great.

The volume of all parts of a mould should be found by careful measurement and calculation, and thus the proper amount of iron can be secured.

Often, even if moulders are good mathematicians, they will, to save a little extra labor in calculating, pour their moulds by *guess-work*, and sometimes find to their sorrow they have been *deceived.* The following *tables* and *rules* are given to assist the moulder in this branch of his trade.

The decimals or fractions of pounds obtained *are not given,* since in *practice* castings can seldom be found to weigh *exactly* what the calculations call for, and the less figures a table contains the easier will they be understood.

To find the weight of square or oblong plates one inch in thickness, multiply the *length by the breadth.* Then multiply the area in cubic inches thus obtained by the decimal .2607 (the weight of a cubic inch of cast iron), which gives the weight of plate in *pounds.* Example :

To find the weight of a square plate 12 inches on the side and one inch thick.

	12″
	12
Cubic inches in plate, -	144
	.2607
	1008
	8640
	288
Weight of plate in pounds and decimals,	37.5408 = $37\frac{54}{100}$ lbs.

To find the weight of round, square, or oblong plates, having oblong, square, or round holes in them, subtract the volume of the hole from the volume of the plate, which, multiplied by .2607, will give the weight in pounds. Sometimes, by referring to the *tables* of weights of square or round plates, the weight of a plate the size of the hole is found, which, subtracted from the weight of a solid plate the size of the outside diameter or square, gives the weight of the ring or plate one inch thick, having a square or round hole.

Knowing the weight of any square or round plates one inch in thickness, it is then very easy to obtain the weight of thicker plates, by simply multiplying the weight by the increased thickness. For instance, if the plate is one and a quarter inch in thickness, the weight will be one fourth more; or, if one and a half inch in thickness, will be one half heavier than the weights given in the tables. But if only three quarters of an inch thickness, here the plate will be one quarter lighter in weight.

In order to test correctness of tables, and obtain the decimals if wanted, the rules and examples are given as shown.

TABLE I.

LENGTH OF SIDE.	FOR SQUARE PLATES 1″ THICK.	WEIGHTS.
12 inches.	For square plates 1″ thick.	37½ lbs.
13 “	“ “ “	44 “
14 “	“ “	“
15		“
16		“
17		“
18		
19		
20		
21		
22		
23		
24		
25		
26		
27		
28		
29		
30		
31		
32		
33		
34		
35		
36		
37		
38		
39		
40 “	“ “ “	417 “

TABLE I.—*Continued.*

LENGTH OF SIDE.	FOR SQUARE PLATES 1″ THICK.	WEIGHTS.
41 inches.	For square plates 1″ thick.	lbs.
42 “	“ “ “	“
43 “	“ “ “	“
44		
45		
46		
47		
48		
49		
50	“	
51	“	
52	“	
53		
54		
55		
56		
57		
58		
59		
60		
61	“	
62	“	
63		
64		
65		
66		
67		
68		
69	“	

TABLE I.—*Continued.*

LENGTH OF SIDE.	FOR SQUARE PLATES 1″ THICK.	WEIGHTS.
70 inches.	For square plates 1″ thick.	lbs.
71 "	" " "	"
72 "	" " "	"
73		
74		
75		
76		
77		
78		
79		
80		
81		
82		
83		
84		
85		
86		
87		
88	"	
89	"	
90	"	
91		
92		
93		
94		
95		
96		
97		
98 "	" "	

TABLE I.—*Continued.*

LENGTH OF SIDE.	FOR SQUARE PLATES 1″ THICK.	WEIGHTS.
99 inches.	For square plates 1″ thick.	lbs.
100 "	" " "	"
101 "	" " "	"
102		
103		
104		
105		
106		
107		
108		
109		
110		
111		
112		
113		
114		
115		
116		
117		
118	"	
119	"	
120	"	
121		
122		
123		
124		
125		
126		
127 "	" "	"

TABLE I.—*Continued.*

LENGTH OF SIDE.	FOR SQUARE PLATES 1″ THICK.	WEIGHTS.
128 inches.	For square plates 1″ thick.	lbs.
129 "	" "	"
130 "	" "	"
131		
132		
133		
134		
135		
136		
137		
138		
139		
140		
141		
142		
143		
144		

To find the weight of round cast-iron plates one inch in thickness.

Square the diameter of plate, and multiply by decimal 7854, which will give area in square inches, which, multiplied by the decimal .2607, will give the weight in pounds.

Example : To find the weight of a round plate 12″ diameter and one inch thick.

Diameter of plate,	12″
	12
	24
	12
Square of diameter,	144
	.7854
	576
	720
	1152
	1008
Area in square inches,	113.0976
	.2607
	7916832
	67858560
	2261952
Weight of plate in pounds and decimals,	29.48454432 = $29\frac{48}{100}$ lbs.

The following table gives the weight of round cast-iron plates, from 12 inches in diameter to 144 inches, the thickness being 1 inch.

TABLE· II.

DIAMETER IN INCHES.	FOR ROUND PLATES 1″ THICK.	WEIGHTS.
12 inches.	For round plates 1″ thick.	lbs.
13 "	" " "	"
14 "	" " "	"
15		
16		
17		
18	" " "	66 "

TABLE II.—*Continued.*

DIAMETER IN INCHES.	FOR ROUND PLATES 1″ THICK.	WEIGHTS.
19 inches.	For round plates 1″ thick.	lbs.
20 "	" " "	"
21 "	" " "	"
22		
23		
24		
25		
26		
27		
28		
29		
30		
31		
32		
33		
34		
35		
36		
37		
38		
39		
40		
41		
42		
43		
44		
45		
46		
47		

TABLE II.—*Continued.*

DIAMETER IN INCHES.	FOR ROUND PLATES 1″ THICK.	WEIGHTS.
48 inches.	For round plates 1″ thick.	lbs.
49	" " "	"
50	" " "	"
51		
52		
53		
54		
55		
56		
57		
58		
59		
60		
61		
62		
63		
64		
65		
66		
67		
68		
69		975
70		
71		
72		
73		
74		
75		
76 "	" " "	"

TABLE II.—*Continued.*

DIAMETER IN INCHES.	FOR ROUND PLATES 1″ THICK.	WEIGHTS.
77 inches.	For round plates 1″ thick.	lbs.
78 "	" " "	
79 "	" " "	
80		
81		
82		
83		
84		
85		
86		
87		
88		
89		
90		
91		
92		
93		
94		
95		
96		
97		
98		
99		
100		
101		
102		
103		
104		
105		

TABLE II.—*Continued.*

DIAMETER IN INCHES	FOR ROUND PLATES 1″ THICK.	WEIGHTS.
106 inches.	For round plates 1″ thick.	lbs.
107 "	" " "	
108 "	" " "	
109	"	
110		
111		
112		
113		
114		
115		
116		
117		
118		
119		
120		
121		
122		
123		
124		
125	"	
126	"	
127		
128		
129		
130		
131		
132		
133		
134		
135		

TABLE II.—*Continued.*

DIAMETER IN INCHES.	FOR ROUND PLATES 1″ THICK.	WEIGHTS.
136 inches.	For round plates 1″ thick.	lbs.
137 "	" " "	"
138 "	" " "	"
139	" "	
140		
141		
142		
143		
144	"	"

To find the weight of cast-iron balls.

Multiply the cube of the diameter in inches by .1365,* and the product is the weight in pounds.—*Haswell.*

Example: To find the weight of a ball 12″ in diameter.

Diameter of ball,	12″	
	12	
	——	1728
Square of diameter,	144	.1365
	12	——
	——	8640
	288	10368
	144	5184
Cube of the diameter	——	1728
in inches,	1728	——

Weight of ball in pounds and decimals, } 235.8720 = 235 $\frac{87}{100}$ lbs.

To find the weight of a hollow ball. Take from the table the weight given for ball having the same outside diameter, and subtract from this the weight given for a ball of the same inside diameter; or multiply the difference of the cubes of the exterior and interior diameter in inches by .1365.

* The volume of a ball can be found by multiplying the cube of the diameter by .5236. The product multiplied by .2607 (the weight of a cubic inch of cast iron) will give the weight in pounds.

TABLE III.

TABLES FOR THE WEIGHT OF BALLS HAVING DIAMETERS FROM 3 INCHES TO 60 INCHES.

DIAMETER IN INCHES.	FOR BALLS 3 INCHES TO 29 INCHES.	WEIGHTS.
3 inches.	Solid balls 3 to 29 inches.	$3\frac{1}{2}$ lbs.
4 "	" " "	$8\frac{3}{4}$ "
5	.. .. ..	17 "
6		$29\frac{1}{2}$ "
7		47 "
8		
9		
10		
11		
12		
13		
14		
15		
16		
17		
18		
19		
20		
21		
22		
23		
24		
25		
26		
27		
28		
29	"	"

TABLE III.—*Continued.*

DIAMETER IN INCHES.	FOR BALLS 30 INCHES TO 60 INCHES.	WEIGHTS.
30 inches.	Solid balls 30 to 60 inches.	3686 lbs.
31 "	" "	4067 "
32 "	" " "	4473 "
33	"	4904
34	"	5365
35		5853
36		6369
37		6914
38		7490
39		
40		
41		
42		"
43		"
44		"
45		"
46		"
47		"
48		"
49		"
50		"
51		"
52		"
53		"
54		"
55		"
56		"
57		"
58		"
59		"
60		"

To find the weight of cast-iron pipes or cylinders.

Find the inside area of a pipe or cylinder by multiplying the square of the inside diameter by .7854, then find the outside area by multiplying the square of the outside diameter by .7854; subtract the *former* from the *latter*, and the product is the area in inches, which, multiplied by .2607 (the weight of a cubic inch of cast iron), gives the weight in pounds for one inch of length.

This product, multiplied by the length in inches, will give the weight.

Example: To find the weight of a pipe or cylinder having an inside diameter of $12\frac{1}{2}''$ and $\frac{3}{4}''$ inch thickness, and $12''$ long.

Outside area,	153.938
Inside area,	122.719
Area of circular ring,	31.219
	.2607
	218533
	1873140
	62438
Weight of one inch long,	8.1387933
	12
Weight of twelve inches long,	97.6655196 = $97\frac{67}{100}$ lbs.

TABLE IV.

TABLE FOR THE WEIGHT OF CAST-IRON PIPES OR CYLINDERS ONE FOOT LONG, VARYING FROM 6 INCHES TO 120 INCHES IN DIAMETER, AND ONE AND TWO INCHES IN THICKNESS.

DIAMETER OF CORE.	WEIGHT OF, 1 INCH THICK.	WEIGHT OF, 2 INCHES THICK.
6 inches.	lbs.	lbs.
7 "	"	"
8 "	"	"
9		
10		
11		
12		
13		
14		
15		
16		
17		
18		
19		
20		
21		
22		
23		
24		
25		
26		
27		
28		
29		
30 "	305 "	629

TABLE IV.—*Continued.*

DIAMETER OF CORE.	WEIGHT OF, 1 INCH THICK.	WEIGHT OF, 2 INCHES THICK.
31 inches.	lbs.	649 lbs.
32 "	"	668 "
33 "	"	688 "
34		708
35		727
36		
37		
38		
39		
40		
41		
42		
43		
44		
45		
46		
47		
48		
49		
50		
51		
52		
53		
54		
55		
56		
57		
58	581	

TABLE IV.—*Continued.*

DIAMETER OF CORE.	WEIGHT OF, 1 INCH THICK.	WEIGHT OF, 2 INCHES THICK.
59 inches.	lbs.	lbs.
60 "	"	"
61 "	"	"
62		
63		
64		
65		
66		
67	668	
68	678	
69	688	
70	699	
71	707	
72	717	
73	727	1474
74	737	
75	747	
76	757	
77	767	
78	777	
79	787	
80		
81		
82		
83		
84		
85		
86		

TABLE IV.—*Continued.*

DIAMETER OF CORE.	WEIGHT OF, 1 INCH THICK.	WEIGHT OF, 2 INCHES THICK.
87 inches.	lbs.	lbs.
88 "	"	"
89 "	"	"
90		
91		
92		
93		
94		
95		
96		
97		
98		
99		
100		
101		
102		
103		
104		
105		
106		
107		
108		
109		
110		
111		
112		
113		
114		2280

TABLE IV.—*Continued.*

DIAMETER OF CORE.	WEIGHT OF, 1 INCH THICK.	WEIGHT OF, 2 INCHES THICK.
115 inches.	lbs.	lbs.
116 "	"	"
117 "	"	"
118		
119		
120		

The weights of *square plates* and *round ones,* also *balls* and *cylinders* here given, comprise a set of tables that the author thinks will be found very useful as a means of assisting moulders in ascertaining the amount of iron required to fill such moulds (for *basing runners* and *gates* he, of course, must add to the weights obtained).

There are many different-shaped castings for which no set of tables can be given. and to find the weight of such the moulder will require special calculations. It is not necessary, in all cases, to take every *crook* or *projection* into special account, as it does not require any great ability to get an average, or to determine what the size of a mould or pattern would be if its irregular *projections* were all leveled or the holes filled up. Thus the size of the mould or pattern would come into plain and even surfaces. It is then an easy matter to obtain the volume or number of cubic feet or inches contained in a mould or pattern ; knowing which we can soon know what amount of iron will be required.

To compute the weight of any *shaped castings,* find the number of cubic inches in the piece, then multiply by any of the decimals given below, and the product will give the weight in pounds approximately.

To ascertain the weights of castings by weighing solid wooden patterns, multiply the weight of pine patterns by sixteen, those of hard wood by twelve, and these products will be an approximation to the weight in iron.

The decimal .2607, the weight for a cubic inch of cast iron, which is here used as a multiplier, is taken from HASWELL. There are two other decimals, .26 and .263, which are very often used in place of .2607, and by using them less figures are required.

To figure on the safe side, as in the case of loam moulds or green sand moulds that are liable to strain much, and also for hard iron, the decimal .263 is the best to use.

For ordinary moulds the decimal .26, used as a multiplier of volumes or areas in inches, will be found to give sufficiently close answers.

Further information upon this subject of figuring for the weights of castings will be found on p. 247, Vol. II.

THE END.

INDEX.

SHORT-TITLE CATALOGUE

OF THE

PUBLICATIONS

OF

JOHN WILEY & SONS,

NEW YORK.

LONDON: CHAPMAN & HALL, LIMITED.

ARRANGED UNDER SUBJECTS.

Descriptive circulars sent on application.
Books marked with an astei isk are sold at *net* prices only.
All books are bound in cloth unless otherwise stated.

AGRICULTURE.

CATTLE FEEDING—DAIRY PRACTICE—DISEASES OF ANIMALS—GARDENING, ETC.

Armsby's Manual of Cattle Feeding....................12mo, $1 75
Downing's Fruit and Fruit Trees..........................8vo, 5 00
Grotenfelt's The Principles of Modern Dairy Practice. (Woll.) 12mo,
Kemp's Landscape Gardening.... ....12mo,
Maynard's Landscape Gardening........................12mo,
Steel's Treatise on the Diseases of the Dog................8vo,
" Treatise on the Diseases of the Ox..................8vo,
Stockbridge's Rocks and Soils.... . ...8vo,
Woll's Handbook for Farmers and Dairymen............12mo,

ARCHITECTURE.

BUILDING—CARPENTRY—STAIRS—VENTILATION—LAW, ETC.

Berg's Buildings and Structures of American Railroads.....4to, 7 50
Birkmire's American Theatres—Planning and Construction.8vo, 3 00
" Architectural Iron and Steel...... ....8vo, 3 50
" Compound Riveted Girders....................8vo, 2 00
" Skeleton Construction in Buildings 8vo, 3 00

Birkmire's Planning and Construction of High Office Buildings. 8vo,
Briggs' Modern Am. School Building....................8vo,
Carpenter's Heating and Ventilating of Buildings..........8vo,
Freitag's Architectural Engineering.... .. ..8vo,
"   The Fireproofing of Steel Buildings.......... ...8vo,
Gerhard's Sanitary House Inspection...................16mo,
"   Theatre Fires and Panics.....................12mo,
Hatfield's American House Carpenter....................8vo,
Holly's Carpenter and Joiner.. ....18mo,
Kidder's Architect and Builder's Pocket-book...16mo, morocco, 4
Merrill's Stones for Building and Decoration..............8vo,
Monckton's Stair Building—Wood, Iron, and Stone........4to, 4
Wait's Engineering and Architectural Jurisprudence.......8vo, 6
Sheep, 6
Worcester's Small Hospitals—Establishment and Maintenance, including Atkinson's Suggestions for Hospital Architecture... ....12mo, 1 25
World's Columbian Exposition of 1893.............Large 4to, 2 50

ARMY, NAVY, Etc.

Military Engineering—Ordnance—Law, Etc.

* Bruff's Ordnance and Gunnery........................8vo,
Chase's Screw Propellers... ....8vo,
Cronkhite's Gunnery for Non-com. Officers.....32mo, morocco,
* Davis's Treatise on Military Law.......................8vo,
Sheep,
* "   Elements of Law................................8vo,
De Brack's Cavalry Outpost Duties. (Carr.)....32mo, morocco,
Dietz's Soldier's First Aid.....................16mo, morocco,
* Dredge's Modern French Artillery....Large 4to, half morocco,
"   Record of the Transportation Exhibits Building, World's Columbian Exposition of 1893..4to, half morocco, 10 00
Durand's Resistance and Propulsion of Ships.............8vo, 5 00
Dyer's Light Artillery.................................12mo, 3 00
Hoff's Naval Tactics....................................8vo, 1 50
* Ingalls's Ballistic Tables...............................8vo, 1 50

BOTANY.

Gardening for Ladies, Etc.

Baldwin's Orchids of New England................Small 8vo, $1 50
Thomé's Structural Botany.... ....16mo, 2 25
Westermaier's General Botany. (Schneider.).............8vo, 2 00

BRIDGES, ROOFS, Etc.

Cantilever—Draw—Highway—Suspension.

(*See also* Engineering, p. 8.)

Boller's Highway Bridges..................................8vo,
* " The Thames River Bridge.................4to, paper,
Burr's Stresses in Bridges....................................8vo,
Crehore's Mechanics of the Girder.........................8vo,
Dredge's Thames Bridges...... ....7 parts, per part,
Du Bois's Stresses in Framed Structures.............Small 4to,
Foster's Wooden Trestle Bridges.......................... ....4to,
Greene's Arches in Wood, etc...............................8vo,
" Bridge Trusses...8vo,
" Roof Trusses...8vo,
Howe's Treatise on Arches 8vo,
Johnson's Modern Framed Structures.............Small 4to,
Merriman & Jacoby's Text-book of Roofs and Bridges.
Part I., Stresses..... ....8vo, 2 50
Merriman & Jacoby's Text-book of Roofs and Bridges.
Part II., Graphic Statics 8vo, 2 50
Merriman & Jacoby's Text-book of Roofs and Bridges.
Part III., Bridge Design..............................8vo, 2 50
Merriman & Jacoby's Text-book of Roofs and Bridges.
Part IV., Continuous, Draw, Cantilever, Suspension, and
Arched Bridges..8vo, 2 50
* Morison's The Memphis Bridge..................Oblong 4to, 10 00
Waddell's Iron Highway Bridges.... ....8vo, 4 00
" De Pontibus (a Pocket-book for Bridge Engineers).
16mo, morocco, 3 00
Wood's Construction of Bridges and Roofs................8vo, 2 00
Wright's Designing of Draw Spans. Parts I. and II..8vo, each 2 50
" " " " " Complete.. .8vo, 3 50

CHEMISTRY—BIOLOGY—PHARMACY.

QUALITATIVE—QUANTITATIVE—ORGANIC—INORGANIC, ETC.

Adriance's Laboratory Calculations....................12mo, $1 25
Allen's Tables for Iron Analysis..........................8vo, 3 00
Austen's Notes for Chemical Students..................12mo, 1 50
Bolton's Student's Guide in Quantitative Analysis.........8vo, 1 50
Boltwood's Elementary Electro Chemistry ....(*In the press.*)
Classen's Analysis by Electrolysis. (Herrick and Boltwood.).8vo, 3 00
Cohn's Indicators and Test-papers........................12mo 2 00
Crafts's Qualitative Analysis. (Schaeffer.)..............12mo, 1 50
Davenport's Statistical Methods with Special Reference to Biological Variations......................12mo, morocco, 1 25
Drechsel's Chemical Reactions. (Merrill.)...............12mo, 1 25
Fresenius's Quantitative Chémical Analysis. (Allen.).......8vo, 6 00
" Qualitative " " (Johnson.).....8vo, 3 00
" " " " (Wells.) Trans. 16th German Edition..............................8vo,
Fuertes's Water and Public Health......................12mo,
Gill's Gas and Fuel Analysis............................12mo,
Hammarsten's Physiological Chemistry. (Mandel.)........8vo,
Helm's Principles of Mathematical Chemistry. (Morgan).12mo,
Ladd's Quantitative Chemical Analysis..................12mo,
Landauer's Spectrum Analysis. (Tingle.)................8vo,
Löb's Electrolysis and Electrosynthesis of Organic Compounds. (Lorenz.)..................................12mo,
Mandel's Bio-chemical Laboratory......................12mo,
Mason's Water-supply...................................8vo,
" Examination of Water..........................12mo,
Meyer's Radicles in Carbon Compounds. (Tingle.)........12mo,
Miller's Chemical Physics.................................8vo,
Mixter's Elementary Text-book of Chemistry............12mo,
Morgan's The Theory of Solutions and its Results.......12mo,
" Elements of Physical Chemistry.................12mo,
Nichols's Water-supply (Chemical and Sanitary)...........8vo,
O'Brine's Laboratory Guide to Chemical Analysis..........8vo,
Perkins's Qualitative Analysis.........................12mo,
Pinner's Organic Chemistry. (Austen.)..................12mo,

Poole's Calorific Power of Fuels 8vo,	$3 00
Ricketts and Russell's Notes on Inorganic Chemistry (Non-metallic)..... . ... Oblong 8vo, morocco,	75
Ruddiman's Incompatibilities in Prescriptions............. 8vo,	2 00
Schimpf's Volumetric Analysis.......... ....12mo,	2 50
Spencer's Sugar Manufacturer's Handbook.....16mo, morocco,	2 00
Handbook for Chemists of Beet Sugar Houses. 16mo, morocco,	3 00
Stockbridge's Rocks and Soils........................ ....8vo,	2 50
* Tillman's Descriptive General Chemistry................8vo,	3 00
Van Deventer's Physical Chemistry for Beginners. (Boltwood.) 12mo,	1 50
Wells's Inorganic Qualitative Analysis.......... ....12mo,	1 50
" Laboratory Guide in Qualitative Chemical Analysis. 8vo,	1 50
Whipple's Microscopy of Drinking-water..................8vo,	3 50
Wiechmann's Chemical Lecture Notes.....................12mo,	3 00
Sugar Analysis........... ....Small 8vo,	2 50
Wulling's Inorganic Phar. and Med. Chemistry..........12mo,	2 00

DRAWING.

ELEMENTARY—GEOMETRICAL—MECHANICAL—TOPOGRAPHICAL.

Hill's Shades and Shadows and Perspective...............8vo,

MacCord's Descriptive Geometry.........................8vo,

" Kinematics.......................................8vo,

" Mechanical Drawing.........................8vo,

Mahan's Industrial Drawing. (Thompson.)........2 vols., 8vo,

Reed's Topographical Drawing. (H. A.)..................4to,

Reid's A Course in Mechanical Drawing..................8vo.

" Mechanical Drawing and Elementary Machine Design. 8vo. (*In the press.*)

Smith's Topographical Drawing. (Macmillan.)...........

Warren's Descriptive Geometry.................. .,

" Drafting Instruments.............. ....

" Free-hand Drawing.... ....

" Linear Perspective........ ....

" Machine Construction ....

Warren's Plane Problems..............................12mo, $1 25

" Primary Geometry.........................12mo

" Problems and Theorems......................... 2

" Projection Drawing 1

Warren's Shades and Shadows........................... 3

" Stereotomy—Stone-cutting.... 2

Whelpley's Letter Engraving ....

ELECTRICITY AND MAGNETISM.

ILLUMINATION—BATTERIES—PHYSICS—RAILWAYS.

Anthony and Brackett's Text-book of Physics. (Magie.) Small 8vo,

Anthony's Theory of Electrical Measurements...........12mo,

Barker's Deep-sea Soundings.........................8vo,

Benjamin's Voltaic Cell...............................8vo,

" History of Electricity........................8vo,

Classen's Analysis by Electrolysis. (Herrick and Boltwood.) 8vo,

Crehore and Squier's Experiments with a New Polarizing Photo-Chronograph..8vo, 3 00

Dawson's Electric Railways and Tramways. Small, 4to, half morocco,

* Dredge's Electric Illuminations....2 vols., 4to, half morocco,

" " " Vol. II..................4to,

Gilbert's De magnete. (Mottelay.).......................8vo,

Holman's Precision of Measurements......................8vo,

" Telescope-mirror-scale Method...........Large 8vo,

Löb's Electrolysis and Electrosynthesis of Organic Compounds. (Lorenz.)..12mo,

*Michie's Wave Motion Relating to Sound and Light.......8vo, 4

Morgan's The Theory of Solutions and its Results........12mo,

Niaudet's Electric Batteries. (Fishback.)............ ..12mo,

Pratt and Alden's Street-railway Road-beds...............8vo,

Reagan's Steam and Electric Locomotives...............12mo,

Thurston's Stationary Steam Engines for Electric Lighting Purposes.......................... ....8vo, 2 50

*Tillman's Heat............... ... .8vo, 1 50

ENGINEERING.

CIVIL—MECHANICAL—SANITARY, ETC.

(*See also* BRIDGES, p. 4; HYDRAULICS, p. 9; MATERIALS OF ENGINEERING, p. 10; MECHANICS AND MACHINERY, p. 12; STEAM ENGINES AND BOILERS, p. 14.)

Baker's Masonry Construction.......................... 8vo,
" Surveying Instruments.........................12mo,
Black's U. S. Public Works......................Oblong 4to,
Brooks's Street-railway Location...... ....16mo, morocco,
Butts's Civil Engineers' Field Book............16mo, morocco,
Byrne's Highway Construction............................8vo,
" Inspection of Materials and Workmanship.......16mo,
Carpenter's Experimental Engineering 8vo,
Church's Mechanics of Engineering—Solids and Fluids....8vo,
" Notes and Examples in Mechanics...............8vo,
Crandall's Earthwork Tables 8vo,
" The Transition Curve...............16mo, morocco,
*Dredge's Penn. Railroad Construction, etc. Large 4to, half morocco,
* Drinker's Tunnelling.....................4to, half morocco,
Eissler's Explosives—Nitroglycerine and Dynamite....... 8vo,
Folwell's Sewerage.......................................8vo,
Fowler's Coffer-dam Process for Piers................. . 8vo.
Gerhard's Sanitary House Inspection...................12mo,
Godwin's Railroad Engineer's Field-book......16mo, morocco,
Gore's Elements of Geodesy....... ....8vo,
Howard's Transition Curve Field-book.........16mo, morocco,
Howe's Retaining Walls (New Edition.).................12mo,
Hudson's Excavation Tables. Vol. II.................. 8vo,
Hutton's Mechanical Engineering of Power Plants........8vo,
" Heat and Heat Engines.........................8vo,
Johnson's Materials of Construction...............Large 8vo,
" Theory and Practice of Surveying........Small 8vo,
Kent's Mechanical Engineer's Pocket-book.....16mo, morocco,
Kiersted's Sewage Disposal..... ....12mo,
Mahan's Civil Engineering. (Wood.)....................8vo,
Merriman and Brook's Handbook for Surveyors....16mo, mor.,
Merriman's Precise Surveying and Geodesy...............8vo,
" Retaining Walls and Masonry Dams..........8vo,
" Sanitary Engineering.........................8vo,
Nagle's Manual for Railroad Engineers........16mo, morocco,
Ogden's Sewer Design.... ....12mo,
Patton's Civil Engineering....... 8vo, half morocco,

Patton's Foundations..........8vo, $5

Pratt and Alden's Street-railway Road-beds..........8vo,

Rockwell's Roads and Pavements in France..........12mo,

Searles's Field Engineering..........16mo, morocco,

" Railroad Spiral..........16mo, morocco,

Siebert and Biggin's Modern Stone Cutting and Masonry...8vo,

Smart's Engineering Laboratory Practice..........12mo,

Smith's Wire Manufacture and Uses..........Small 4to,

Spalding's Roads and Pavements..........12mo,

" Hydraulic Cement..........12mo,

Taylor's Prismoidal Formulas and Earthwork..........8vo,

Thurston's Materials of Construction..........8vo,

* Trautwine's Civil Engineer's Pocket-book....16mo, morocco,

* " Cross-section.......... Sheet,

* " Excavations and Embankments..........8vo,

* " Laying Out Curves..........12mo, morocco,

Waddell's De Pontibus (A Pocket-book for Bridge Engineers).
16mo, morocco, 3 00

Wait's Engineering and Architectural Jurisprudence..........8vo, 6 00
Sheep, 6 50

" Law of Field Operation in Engineering, etc..........8vo.

Warren's Stereotomy—Stone-cutting..........8vo, 2 50

Webb's Engineering Instruments. New Edition. 16mo, morocco, 1 25

Wegmann's Construction of Masonry Dams..........4to, 5 00

Wellington's Location of Railways..........Small 8vo, 5 00

Wheeler's Civil Engineering..........8vo, 4 00

Wolff's Windmill as a Prime Mover..........8vo, 3 00

HYDRAULICS.

WATER-WHEELS—WINDMILLS—SERVICE PIPE—DRAINAGE, ETC.

(*See also* ENGINEERING, p. 8.)

Bazin's Experiments upon the Contraction of the Liquid Vein.
(Trautwine.)..........8vo, 2 00

Bovey's Treatise on Hydraulics..........8vo, 4 00

Coffin's Graphical Solution of Hydraulic Problems..........12mo, 2 50

Ferrel's Treatise on the Winds, Cyclones, and Tornadoes...8vo, 4 00

Folwell's Water Supply Engineering..........8vo, 4 00

Fuertes's Water and Public Health.......... 12mo, 1 50

Ganguillet & Kutter's Flow of Water. (Hering & Trautwine.)
8vo, 4 00

Hazen's Filtration of Public Water Supply..........8vo, 3 00

Herschel's 115 Experiments..........8vo, 2 00

Kiersted's Sewage Disposal............................12mo,
Mason's Water Supply..................................8vo,
" Examination of Water. 12mo,
Merriman's Treatise on Hydraulics........................8vo,
Nichols's Water Supply (Chemical and Sanitary)..........8vo,
Wegmann's Water Supply of the City of New York.......4to,
Weisbach's Hydraulics. (Du Bois.).....................8vo,
Whipple's Microscopy of Drinking Water...............8vo,
Wilson's Irrigation Engineering..........................8vo,
" Hydraulic and Placer Mining..................12mo,
Wolff's Windmill as a Prime Mover......................8vo,
Wood's Theory of Turbines..................................8vo,

MANUFACTURES.

BOILERS—EXPLOSIVES—IRON—STEEL—SUGAR—WOOLLENS, ETC.

Allen's Tables for Iron Analysis..........................8vo,
Beaumont's Woollen and Worsted Manufacture.........12mo,
Bolland's Encyclopædia of Founding Terms...............12mo,
" The Iron Founder.................................12mo,
" " " " Supplement.................12mo,
Bouvier's Handbook on Oil Painting....................12mo,
Eissler's Explosives, Nitroglycerine and Dynamite........8vo, 4
Ford's Boiler Making for Boiler Makers.................18mo,
Metcalfe's Cost of Manufactures 8vo,
Metcalf's Steel—A Manual for Steel Users...............12mo,
* Reisig's Guide to Piece Dyeing.........................8vo,
Spencer's Sugar Manufacturer's Handbook....16mo, morocco,
" Handbook for Chemists of Beet Sugar Houses.
16mo, morocco,
Thurston's Manual of Steam Boilers........................8vo,
Walke's Lectures on Explosives............................8vo,
West's American Foundry Practice......................12mo,
" Moulder's Text-book 12mo,
Wiechmann's Sugar Analysis.................... Small 8vo,
Woodbury's Fire Protection of Mills........................8vo,

MATERIALS OF ENGINEERING.

STRENGTH—ELASTICITY—RESISTANCE, ETC.

(*See also* ENGINEERING, p. 8.)

Baker's Masonry Construction..............................8vo, 5 00
Beardslee and Kent's Strength of Wrought Iron 8vo, 1 50
Bovey's Strength of Materials..............................8vo, 7 50
Burr's Elasticity and Resistance of Materials..............8vo, 5 00

Byrne's Highway Construction....... ....8vo, $5 00
Church's Mechanics of Engineering—Solids and Fluids.....8vo,
Du Bois's Stresses in Framed Structures.............Small 4to,
Johnson's Materials of Construction.....................8vo,
Lanza's Applied Mechanics...............................8vo,
Martens's Testing Materials. (Henning.)..........2 vols., 8vo,
Merrill's Stones for Building and Decoration............8vo,
Merriman's Mechanics of Materials.......................8vo,
"   Strength of Materials.......................12mo,
Patton's Treatise on Foundations........................8vo,
Rockwell's Roads and Pavements in France.............12mo, 25
Spalding's Roads and Pavements........................12mo,
Thurston's Materials of Construction....................8vo,
"   Materials of Engineering...............3 vols., 8vo,
Vol. I, Non-metallic 8vo,
Vol. II., Iron and Steel.........................8vo,
Vol. III., Alloys, Brasses, and Bronzes..............8vo,
Wood's Resistance of Materials...........................8vo,

MATHEMATICS.

Calculus—Geometry—Trigonometry, Etc.

Baker's Elliptic Functions...............................
Barnard's Pyramid Problem...............................
*Bass's Differential Calculus..........................
Briggs's Plane Analytical Geometry..................
Chapman's Theory of Equations........................
Compton's Logarithmic Computations
Davis's Introduction to the Logic of Algebra.............
Halsted's Elements of Geometry.........................
"   Synthetic Geometry..   ....
Johnson's Curve Tracing..............   ....
"   Differential Equations—Ordinary and Partial.
Small 8vo, 3 50
"   Integral Calculus...........................12mo, 1 50
"   "   "

"   Least Squares   1 50
*Ludlow's Logarithmic and Other Tables. (Bass.)   2 00
*   "   Trigonometry with Tables. (Bass.).............   3 00
*Mahan's Descriptive Geometry (Stone Cutting) 1 50
Merriman and Woodward's Higher Mathematics.......... ....   5 00
Merriman's Method of Least Squares 2 00
Rice and Johnson's Differential and Integral Calculus,
2 vols. in 1, small 8vo,   2 50

Rice and Johnson's Differential Calculus............Small 8vo, $3 00
" Abridgment of Differential Calculus.
Small 8vo, 1
Totten's Metrology.............. ....8vo, 2
Warren's Descriptive Geometry................. 2 vols., 8vo, 3
" Drafting Instruments......................... 12mo, 1
" Free-hand Drawing..... ....12mo, 1
" Linear Perspective............................12mo, 1
" Primary Geometry..........................12mo,
" Plane Problems..................................12mo, 1
" Problems and Theorems.........................8vo, 2
" Projection Drawing..........................12mo, 1
Wood's Co-ordinate Geometry...............................8vo, 2
" Trigonometry..... ....12mo, 1
Woolf's Descriptive Geometry.....................Large 8vo, 3

MECHANICS—MACHINERY.

TEXT-BOOKS AND PRACTICAL WORKS.

(*See also* ENGINEERING, p. 8.)

Baldwin's Steam Heating for Buildings 12mo,
Barr's Kinematics of Machinery...............................8vo,
Benjamin's Wrinkles and Recipes 12mo,
Chordal's Letters to Mechanics........................ 12mo,
Church's Mechanics of Engineering.... .. ..8vo,
" Notes and Examples in Mechanics..............8vo,
Crehore's Mechanics of the Girder.........................8vo,
Cromwell's Belts and Pulleys......................... 12mo,
" Toothed Gearing...... ....12mo,
Compton's First Lessons in Metal Working.............12mo,
Compton and De Groodt's Speed Lathe..................12mo,
Dana's Elementary Mechanics 12mo,
Dingey's Machinery Pattern Making......................12mo,
Dredge's Trans. Exhibits Building, World Exposition.
Large 4to, half morocco,
Du Bois's Mechanics. Vol. I., Kinematics 8vo,
" " Vol. II., Statics..................8vo,
" " Vol. III., Kinetics...... ....8vo,
Fitzgerald's Boston Machinist..................... 18mo,
Flather's Dynamometers............. ....12mo,
" Rope Driving.............................12mo,
Hall's Car Lubrication...................................12mo,
Holly's Saw Filing 18mo,
Johnson's Theoretical Mechanics. An Elementary Treatise.
(*In the press.*)
Jones's Machine Design. Part I., Kinematics.............8vo, 1 50

Jones's Machine Design. Part II., Strength and Proportion of Machine Parts.......................................8vo, $3 00
Lanza's Applied Mechanics..................................8vo,
MacCord's Kinematics..8vo,
Merriman's Mechanics of Materials.........................8vo,
Metcalfe's Cost of Manufactures............................8vo,
*Michie's Analytical Mechanics..............................8vo,
Richards's Compressed Air...................................12mo,
Robinson's Principles of Mechanism.........................8vo,
Smith's Press-working of Metals..............................8vo, 3
Thurston's Friction and Lost Work...........................8vo,
" The Animal as a Machine....................12mo,
Warren's Machine Construction........................2 vols., 8vo,
Weisbach's Hydraulics and Hydraulic Motors. (Du Bois.)..8vo,
" Mechanics of Engineering. Vol. III., Part I., Sec. I. (Klein.)..8vo, 5 00
Weisbach's Mechanics of Engineering. Vol. III., Part I., Sec. II. (Klein.)..8vo, 5 00
Weisbach's Steam Engines. (Du Bois.)........................8vo, 5 00
Wood's Analytical Mechanics..................................8vo, 3 00
" Elementary Mechanics..............................12mo, 1 25
" " " Supplement and Key.....12mo, 1 25

METALLURGY.

IRON—GOLD—SILVER—ALLOYS, ETC.

Allen's Tables for Iron Analysis..............................
Egleston's Gold and Mercury....................
" Metallurgy of Silver....................
* Kerl's Metallurgy—Copper and Iron....................
* " " Steel, Fuel, etc....................
Kunhardt's Ore Dressing in Europe....................
Metcalf's Steel—A Manual for Steel Users
O'Driscoll's Treatment of Gold Ores....................
Thurston's Iron and Steel..............................
" Alloys..............................
Wilson's Cyanide Processes....................

MINERALOGY AND MINING.

MINE ACCIDENTS—VENTILATION—ORE DRESSING, ETC.

Barringer's Minerals of Commercial Value....Oblong morocco, 2 50
Beard's Ventilation of Mines..................................12mo, 2 50
Boyd's Resources of South Western Virginia..............8vo, 3 00
" Map of South Western Virginia.....Pocket-book form, 2 00
Brush and Penfield's Determinative Mineralogy. New Ed. 8vo, 4 00

Chester's Catalogue of Minerals...... ....8vo,
" " " ....Paper,
" Dictionary of the Names of Minerals.............8vo,
Dana's American Localities of Minerals.............Large 8vo,
" Descriptive Mineralogy (E. S.) Large 8vo. half morocco,
" First Appendix to System of Mineralogy. .. Large 8vo,
" Mineralogy and Petrography. (J. D.)...........12mo,
" Minerals and How to Study Them. (E. S.). 12mo,
" Text-book of Mineralogy. (E. S.)...New Edition. 8vo,
* Drinker's Tunnelling, Explosives, Compounds, and Rock Drills. 4to, half morocco,
Egleston's Catalogue of Minerals and Synonyms...........8vo,
Eissler's Explosives—Nitroglycerine and Dynamite........8vo,
Hussak's Rock-forming Minerals. (Smith.).........Small 8vo,
Ihlseng's Manual of Mining.. .. ..8vo,
Kunhardt's Ore Dressing in Europe.....................8vo,
O'Driscoll's Treatment of Gold Ores....................8vo,
* Penfield's Record of Mineral Tests...... ....Paper, 8vo,
Rosenbusch's Microscopical Physiography of Minerals and Rocks. (Iddings.)...................... .. .8vo,
Sawyer's Accidents in Mines......................Large 8vo,
Stockbridge's Rocks and Soils..........................8vo,
Walke's Lectures on Explosives.........................8vo,
Williams's Lithology..... ....8vo,
Wilson's Mine Ventilation 12mo,
" Hydraulic and Placer Mining.... ....12mo,

STEAM AND ELECTRICAL ENGINES, BOILERS, Etc.

STATIONARY—MARINE—LOCOMOTIVE—GAS ENGINES, ETC.

(*See also* ENGINEERING, p. 8.)

Baldwin's Steam Heating for Buildings.................12mo,
Clerk's Gas Engine ....Small 8vo,
Ford's Boiler Making for Boiler Makers.................18mo,
Hemenway's Indicator Practice........................12mo,
Hoadley's Warm-blast Furnace........... ....8vo,
Kneass's Practice and Theory of the Injector 8vo,
MacCord's Slide Valve..................................8vo,
Meyer's Modern Locomotive Construction.................
Peabody and Miller's Steam-boilers...................... ...
Peabody's Tables of Saturated Steam. ....
" Thermodynamics of the Steam Engine.........
" Valve Gears for the Steam Engine.............
Pray's Twenty Years with the Indicator..
Pupin and Osterberg's Thermodynamics......

Reagan's Steam and Electric Locomotives............ ......12mo,	$2 00
Röntgen's Thermodynamics. (Du Bois.)................8vo,	5 00
Sinclair's Locomotive Running.........................12mo,	2 00
Snow's Steam-boiler Practice...... ....8vo.	3 00
Thurston's Boiler Explosions.... ....12mo,	1 50
" Engine and Boiler Trials............................8vo,	5 00
" Manual of the Steam Engine. Part I., Structure and Theory.................................8vo,	6 00
" Manual of the Steam Engine. Part II., Design, Construction, and Operation...............8vo,	6 00
2 parts,	10 00
Thurston's Philosophy of the Steam Engine.............12mo,	75
" Reflection on the Motive Power of Heat. (Carnot.) 12mo,	
" Stationary Steam Engines.................. .. 8vo,	
" Steam-boiler Construction and Operation...... 8vo,	
Spangler's Valve Gears....................................8vo,	
Weisbach's Steam Engine. (Du Bois.)..................8vo,	
Whitham's Constructive Steam Engineering...............8vo,	
" Steam-engine Design.... ....8vo,	
Wilson's Steam Boilers. (Flather.)....................12mo,	
Wood's Thermodynamics, Heat Motors, etc..............8vo,	

TABLES, WEIGHTS, AND MEASURES.

For Actuaries, Chemists, Engineers, Mechanics—Metric Tables, Etc.

Adriance's Laboratory Calculations....................	1
Allen's Tables for Iron Analysis	3
Bixby's Graphical Computing Tables........	
Compton's Logarithms.	1
Crandall's Railway and Earthwork Tables.....	1
Egleston's Weights and Measures......................	
Fisher's Table of Cubic Yards....................	
Hudson's Excavation Tables. Vol. II.....	1
Johnson's Stadia and Earthwork Tables	1
Ludlow's Logarithmic and Other Tables. (Bass.)	2
Totten's Metrology....	2

VENTILATION.

Steam Heating—House Inspection—Mine Ventilation.

Baldwin's Steam Heating.............................. 12mo,	2 50
Beard's Ventilation of Mines........................12mo,	2 50
Carpenter's Heating and Ventilating of Buildings..........8vo,	3 00
Gerhard's Sanitary House Inspection..................12mo,	1 00
Wilson's Mine Ventilation...............................12mo,	1 25

MISCELLANEOUS PUBLICATIONS.

Alcott's Gems, Sentiment, Language..............Gilt edges,	
Davis's Elements of Law...................................8vo,	
Emmon's Geological Guide-book of the Rocky Mountains..8vo,	
Ferrel's Treatise on the Winds...........................8vo,	
Haines's Addresses Delivered before the Am. Ry. Assn...12mo,	
Mott's The Fallacy of the Present Theory of Sound..Sq. 16mo,	
Richards's Cost of Living..............................12mo,	
Ricketts's History of Rensselaer Polytechnic Institute.....8vo,	
Rotherham's The New Testament Critically Emphasized. 12mo,	1 50
" The Emphasized New Test. A new translation. Large 8vo,	2 00
Totten's An Important Question in Metrology. 8vo,	2 50
* Wiley's Yosemite, Alaska, and Yellowstone 4to,	3 00

HEBREW AND CHALDEE TEXT-BOOKS.

For Schools and Theological Seminaries.

Gesenius's Hebrew and Chaldee Lexicon to Old Testament. (Tregelles.)...................Small 4to, half morocco,	5 00
Green's Elementary Hebrew Grammar.................12mo,	1 25
" Grammar of the Hebrew Language (New Edition).8vo,	3 00
" Hebrew Chrestomathy..........................8vo,	2 00
Letteris's Hebrew Bible (Massoretic Notes in English). 8vo, arabesque,	2 25

MEDICAL.

Hammarsten's Physiological Chemistry. (Mandel.)........8vo,	4 00
Mott's Composition, Digestibility, and Nutritive Value of Food. Large mounted chart,	1 25
Ruddiman's Incompatibilities in Prescriptions............8vo,	2 00
Steel's Treatise on the Diseases of the Ox.... ....8vo,	6 00
" Treatise on the Diseases of the Dog...............8vo,	3 50
Woodhull's Military Hygiene.......... ....16mo,	1 50
Worcester's Small Hospitals—Establishment and Maintenance, including Atkinson's Suggestions for Hospital Architecture...12mo,	1 25

CPSIA information can be obtained at www.ICGtesting.com
Printed in the USA
LVOW10s2134080816

499569LV00015B/200/P